Environmentalism and Contemporary Heterotopia

Environmentalism and Contemporary Heterotopia

Novel Encounters with Waste

Tom Bowers

LEXINGTON BOOKS
Lanham • Boulder • New York • London

Published by Lexington Books
An imprint of The Rowman & Littlefield Publishing Group, Inc.
4501 Forbes Boulevard, Suite 200, Lanham, Maryland 20706
www.rowman.com

86-90 Paul Street, London EC2A 4NE

Copyright © 2022 by The Rowman & Littlefield Publishing Group, Inc.

All rights reserved. No part of this book may be reproduced in any form or by any electronic or mechanical means, including information storage and retrieval systems, without written permission from the publisher, except by a reviewer who may quote passages in a review.

British Library Cataloguing in Publication Information Available

Library of Congress Cataloging-in-Publication Data

978-1-7936-2297-6
978-1-7936-2299-0
978-1-7936-2298-3

Contents

Introduction	1
1 The Inescapable Presence of Dirty Matter	11
2 The Potential of Dissonance and Heterotopia	33
3 Beyond Sustainability: Relationality, Uncertainty, and the Responsible Posthuman Environmental Public	53
4 Reorientations to Risk	89
5 The Ethics of Agency in a Dirty World	119
Conclusion	149
Bibliography	153
Index	167
About the Author	173

Introduction

This book is about American society's twenty-first-century relation to industrial waste. The exploration undertaken in this book is grounded in the premise that the ubiquity and vibrancy of industrial waste have constituted heterotopia, spaces where the juxtaposition of waste and human activity jars the conceptual apparatus that has guided modern environmentalism. Whereas modern relations to waste have been informed by a consciousness that aims to make waste absent by attempting to materially and psychologically distance the matter from human contact and to regain a sense of cleanliness and order, the heterotopia explored in this book provoke the public consciousness to concede that human encounters with waste are unavoidable realities of the current epoch. With its focus on heterotopia, this book is situated within concerns of space and place, concepts long associated with environmentalism and with modern relations of waste and play. For example, the practice of remediation, whose premises have been consistently challenged, has, nonetheless become a staple of modern waste relations, as spaces contaminated by industrial detritus become transformed into sites of recreation and play once the dirty material is removed and placed elsewhere. In addition, the environmental justice movement has revised the definition of the environment so that ecological care entails not just attending to the human impacts upon spaces of natural beauty but also those spaces in which humans work, live, and play. Through the lens of environmental justice, the spatial nature of the modern relation to waste becomes associated with the economic and racial characteristics of the spaces a public inhabits.

In the twenty-first century, the ubiquity of waste and the resulting novel configurations of dirty matter and human activities generate an opportunity to identify and deliberate upon the associations among a range of newly emplaced human and nonhuman objects and the prospects for dwelling with waste that such unfamiliar arrangements and relations foster. As Foucault's ([1967] 1986) definition establishes, heterotopia are "capable of juxtaposing in a single real place several spaces, several sites that are themselves incompatible" (3). The incongruity fostered by such multiplicity and novel juxtapositions makes heterotopia sites of conceptual provocation given that they are "something like counter-sites, a kind of effectively enacted utopia in which the real sites, all the other real sites that can be found within the culture, are simultaneously represented, contested, and inverted" (2). The incompatibility that results from waste's presence and the potential emergence of new conceptual apparatuses make these distinct spaces valuable lines of inquiry by which to consider the process of social change, particularly in terms of the way these spaces prompt the need for novel perceptions of environmentalism. This book focuses on the possibilities afforded by three specific heterotopia: a space where industrial effluent and industrial practices coincide with the preservation of an endangered species and the tourist activities of encountering this species; a space where a landfill of radioactive waste and the practices of managing such dangerous matter generate a recreational and tourist destination; a space where a speedboat race and cultural festival annually occur on a lake designated as a Superfund site. Through an analysis of these three heterotopia, this book seeks to contribute to an understanding of waste and contemporary environmentalism in three ways. First, it reaffirms the importance of space and place to environmentalism, but argues for the value of heterotopia as a means to recognize and analyze the emergence of novel twenty-first-century spaces and the way that environmentalism must discard familiar habits of thought to respond to and address the incongruity and dissonance that such spaces sponsor. Second, it attempts to direct attention to the quotidian and the way that a publics' environmental consciousness is produced by the capacity to locate oneself within everyday contemporary entanglements with waste. The specialized texts of environmental advocacy, be they literature, film, and nonfiction treatises, serve an important function in challenging prevailing and often destructive practices and rhetorics. But understanding the space for an emerging environmentalism requires locating ways that publics encounter, understand, and act within novel, contemporary spaces of dissonance. Third, it questions the promises of new materialism and the posthuman to generate a renewed environmental ethic, positing instead that relationality may lead to new associations but such material-discursive arrangements, as emergent in a contemporary dirtied world, need not always lead to ethical environmental subjects and practices.

Central to posthuman and new materialist theories is the agency granted to the nonhuman. Waste's material unruliness and vibrancy are evident through its capacity to alter various bodies and through the constraints placed upon human actors to restrain dirty matter's capacity to reconstitute space and the environment. As a result, humans need a perceptual reorientation to a world that is becoming and transforming in unpredictable ways due to waste's ubiquity and vibrancy:

> From our understanding of the boundary between life and death and our everyday work practices to the way we feed ourselves and recreate or procreate, we are finding our environment materially and conceptually reconstituted in ways that pose profound and unprecedented normative questions. In addressing them, we unavoidably find ourselves having to think in new ways about the nature of matter and the matter of nature; about the elements of life, the resilience of the planet, and the distinctiveness of the human." (Coole and Frost 2010, 6)

For posthumanists, the reconfigurations of space brought about by the agency of the nonhuman prompts the potential for a more ethical attunement toward the planet's other-than-human inhabitants and an enhanced commitment to environmental care. As I contend in this book, waste has indeed reshaped the contemporary landscape. But the constitution of the resulting heterotopia may not offer up the potential for such an ethical stance. If the nonhuman is granted agency and the creative capacity to reconfigure the world and to resist the influence of human intentions, then we must grant the possibility that these unpredictable and novel entanglements may not allow for the ethical, environmental frames we desire or have long become familiar with.

The potential for and challenges to a renewed and enhanced environmental ethic invoked by a reconstituted reality are evident in the failed promises of Ulrich Beck's risk society and reflexive modernity. As Beck ([1986] 2011) has argued, the pervasiveness of toxins has transformed the world into risk society, a particular spatiotemporal situation that calls into question the central practices and consciousness of Western modernity, particularly those practices informed by scientific and technological expertise. The resulting skepticism in these epistemologies, what Beck refers to as reflexive modernization, thus leads to the formation of a new political subject, one equipped with ways of knowing and being in the world that extend beyond the objective tenets of science. Beck's thesis has drawn much support, with the significant attention given to trying to authorize alternative forms of knowledge as a means to define risks from toxins as clear evidence. But while such theoretical assertions and alternative epistemologies have been advanced, the American public's encounter and involvement with waste in the second decade of the twenty-first century, even with the emergence of various novel configurations, remains largely informed by practices grounded in modernity, affirming the persistence of the prevailing order that has led to significant environmental destruction. In the twenty-first century, waste may increasingly reveal its capacity to resist human directions, to move on its own accord, and to intra-act with a range of other material to constitute novel material and conceptual spaces that afford the potential to reorient human consciousness and human practices, yet such material articulations have failed to fully revise the modern human subject and modern practices.

The rigidity of modern practices does not, however, suggest that alternative ways of perceiving and being with waste are not possible. Because of their inherent incongruity and disruptiveness and their capacity to reflect, invert, and contest prevailing orders, twenty-first-century heterotopia offer the potential for new practices and novel ways of being in the world of waste to emerge. In speaking of the relation between reconfigured space and rhetoric, Stormer and MacGreavey (2017) contend, "Different materialities set the field of potential and condition diverse rhetorics' emergence from the broader environment. If that environment changes, so too does rhetorical capacity" (19). The rearrangement of material that constitutes the novel spaces and configurations provide the space in which novel discursive arrangements and articulations may emerge. Heterotopia, through the new material-discursive arrangements and rhetorics they engender, hold the possibility of reshaping human performance and the way publics perceive and act within the novel configurations of the current epoch. But the emergence of such subjects and their performances rests upon the capacity to interpret one's emplacement within these spaces, perceptions that entail not just reacknowledging one's associations with dirty matter but also one's novel relations with other human and nonhuman participants. Heterotopia are then first an interpretative problem, a conceptual interruption and disjuncture in need of some hermeneutic resolution. Any response is thus dependent upon one's proficiency to interpret one's emplacement, particularly the novel relations that one finds oneself entangled within. As the case studies used in this book will show, the incongruity generated by the twenty-first-century heterotopia is often resolved by reference to familiar habits of interpretation and common perceptions and practices found in modern environmentalism. But given the multiplicity and incongruous quality of heterotopia, these modernist responses to waste fail to account for not only the increased relationality that marks contemporary heterotopia but also the emerging ethical incongruities that result from the novel and disparate juxtapositions that constitute these spaces. The reliance on and subsequent limits of the familiar perceptions and rhetorics attached to modern environmentalism call up the need to generate a new hermeneutical and rhetorical proficiency, a capacity made possible through a public's engagement with heterotopic spaces of waste. This book thus calls for the need to instill within a public the requisite interpretive capacity to read and respond to the novel material-discursive spaces that are becoming increasingly common, a capacity that affords the opportunity to understand one's emplacement within a dirtied world and the ethics involved in such relations.

As I suggest, heterotopia are a hermeneutic and rhetorical resource, a space in which interpretation and action become essential to resolve the conceptual and even material discord and whose possibilities to offer such reconciliation challenge familiar environmental perceptions and ethics. Yet a reliance on the familiar to resolve the dissonance generated by waste's pervasiveness should not be surprising. The reliance on the familiar rhetoric of the modern can be seen as interpreting the discord as an opportunity to restore the utopian promises assumed through human mastery. This utopian rhetorics stands in contrast to a second familiar interpretation an encounter with waste provokes. If the utopian conjures a perception and praxis led by human mastery and modern enlightenment, the dystopian provokes, with its images of desolate landscapes soiled by waste that cannot be controlled and managed, humanity's incapacities and habits of self-destruction. But while these frames of thought and corresponding rhetorics may often be drawn upon to understand the unfamiliarity of current spaces, they are ill-suited to understand heterotopia and the novel practices of thought and action that such incongruity engender since neither familiar frame provides a credible account of a public's relation to the dirty, contemporary world. Heterotopia thus afford the means to understand the dirty world as an alternative to these common and limited tropes. "Neither materialized utopias, nor spaces of urban entropy, dystopia or socio-spatial vulnerability, heterotopias are not the margin. Spatial heterotopias are exceptions that differ so greatly from all categories that they cannot be fitted and fixed into any rigid taxonomy" (Sohn 2008, 49). Heterotopia thus exhibit a hermeneutical amorphousness, a multiplicity and openness responsive to recent discussions concerning the shortcomings of familiar forms of toxic and environmental discourse.

Scholars tracing the evolution of environmental rhetorics have noted the limitations of a rhetoric promoting ecological utopia, especially discourse that promotes the possibilities of ecological equilibrium and attaining any semblance of a pure, natural order. Simultaneously, scholars have come to accept the limits of the dystopian and the capacity of apocalyptic visions of a doomed planet to promote public action. Attention has since been directed at the in-between, a discourse Buell (2003) describes as "dwelling in crisis." Buell identifies Beck's risk society as a central text within this new rhetoric and sees the rhetorics of dwelling in crisis, as with Beck's risk society, as a means to promote a new political and environmental human subject. As Buell contends, society is facing "a new period in human history. First, society has built up a considerable volume of ecological karma—a legacy of past problems created and not solved. Second, it has been forced to recognize that fact. And third, its powers to affect the environment have grown and its remaining ecological space has shrunk" (178). A discourse that promotes a balance between humans and nature is thus no longer feasible given the prominence of the spaces impacted by human activities and the inability of the human to adequately address the consequences of its actions and cleanse the planet from all dirty matter. Moreover, the end prophesized in apocalyptic tracts has yet to materialize. Instead, we find ourselves situated within spaces in which we continually encounter the past sins of industrial production and consumption. And other than succumbing to the guilt and shame attached to the ecological misdeeds that have diminished any hopes of an earthly Eden and that will lead us toward environmental catastrophe, we are given no other interpretative apparatus by which to live with and relate to dirt's continued existence. Contemporary heterotopia afford an alternative, allowing us to recognize that even with an acknowledged guilt, absolution, and efforts to refrain from detrimental ecological practices, we, nevertheless, find ourselves inescapably emplaced within a modified, dirtied world.

This book thus moves ecological awareness beyond the utopian and the apocalyptic, moving the discourse away from notions of balance and purity and end of the world projections to concerns about how to exist within the present, damaged ecology. "In the process, environmental crisis has been richly reimagined and made tellingly more relevant than ever to present social and political concerns" (Buell 2003, 31). Buell's focus is on the emergence of the novel discourse in contemporary literature, film, and nonfiction works of prose; but the characteristics of the discourse he outlines, as this book proposes, have particular relevance to providing a hermeneutic and performative resource by which to encounter and act within everyday, contaminated space. "For dwelling in crisis means facing the fact that one dwells in a body and in ecosystems, both of which are already subject to considerable degradation, modification, and pressure. No credible refuge from damage to these is at hand" (187). The task thus becomes how to become equipped to dwell and even play within the dirt.

This book embarks on the initial steps of such an inquiry and proceeds through the following sequence of chapters. In chapter 1, I survey the modern, human relation to waste, a relation largely informed by a perceived absence of the dirty matter, as a way to establish the primary assertion that familiar environmental rhetorics and practices are no longer viable in a dirtied world. Absence once played the foundational role in informing environmental perceptions of and relations to waste, an assertion I make evident through reference to a series of material configurations that spurred familiar environmental practices and perceptions. These habits of thought and action, at least through the mid-twentieth century, had a beneficial impact on lessening the presence of the dirty material. After a series of successful efforts, however, these common practices have since failed to prevent waste from becoming increasingly entangled with human encounters and have also failed to prevent the vibrant quality of the dirty matter from generating novel configurations in which humans now find themselves emplaced. Subsequently, the prevailing environmental perceptions and practices, as chapter 1 contends, do not fully account for the posthuman reality and heterotopic spaces that industrial waste has had a role in creating. Therefore, I outline how new materialism provides an important first step to understand and respond to these new configurations in which humans now find themselves entangled.

In chapter 2, I suggest that a new exigence has emerged with respect to the presence of waste, one that calls upon the need to generate an appropriate hermeneutic capacity to interpret and respond to these new spaces and configurations. Exigence is no longer defined by the appearance of dirty matter and the efforts by humans to remove it; instead, exigence now entails acknowledging waste's ever-presence and vibrancy and developing the processes to understand and generate the means by which humans can dwell with the dissonance and disturbance caused by waste's unwillingness to be removed. Developing the needed hermeneutical proficiency appropriate for the contemporary dirty world thus requires the capacity to engage with heterotopia, with configurations marked by dissonance and the out-of-place, where the familiar perceptions and practices that informed human relations to waste are no longer sufficient and the prospects for new perceptions and responses emerge from and through the novel relations and arrangements made available by contemporary, incongruous configurations. I have previously discussed (Bowers 2018) heterotopia in relation to waste and actor-network theory, but in this chapter I extend the conversation by associating dissonant spaces with rhetorical situations. Space, relationality, and rhetoric all become important concepts to generate the perceptions and responses more fitting to the dirty world and help reveal the ethical complexity that emerges through the novel material-discursive configurations.

In chapter 3, I turn to the first of a series of case studies to explore the way that familiar environmental frames of thought seem ill-fitting to contemporary heterotopia and to advance the need to attend to the emergence of novel material-discursive relations and the ethical issues that these arrangements evoke through new perceptual and hermeneutic ways. By establishing links among several sites in which waste is present in the state of Florida, I explore the limits of several familiar habits of environmental thought, including those efforts directed at limiting emissions and other waste products. The fidelity to these efforts, while benefiting future ecologies, fails to fully account for the present, dirtied state of the world and the complex arrangements that have emerged from waste's presence and vibrancy. In other words, the familiar habit of environmental thought offers little guidance as to how humans should encounter dirtied spaces other than to see them as shameful examples of human hubris and inattention to the other-than-human. I also join several other voices to challenge prevailing notions of sustainability, a practice and rhetoric grounded in stasis and the perception that balancing human and nonhuman needs can be attained and maintained by the human capacity to shape the actions of the various actors populating a confined and bordered space. Through my discussion in this chapter, I illustrate the fallacy of such a belief and also explain how familiar patterns of cause and effect and subsequently responsibility for dwelling with waste must be revised. Such common environmental tropes as downwind and downstream used to assign blame and responsibility for the movement of toxins are no longer sufficient to explain the new and unpredictable relations and entanglements with waste that emerge within the range of dispersed configurations that vibrant nonhuman actors are able to constitute. The emergence of such distributed arrangements and the vibrancy of various actors, both human and nonhuman, within these dispersed entanglements to influence the shape of these configurations call attention to the way we assign responsibility to publics and institutions and how we locate and define ecological care.

In chapter 4, I draw from another heterotopic site to further explore the concepts of space, relationality, and ethics within a dirty world. Specifically, I draw from my experience touring the Weldon Spring Site, a former nuclear weapons production location that has since become a cultural heritage and recreational tourist destination. What makes Weldon Spring particularly heterotopic is the public capacity to hike atop and alongside a radioactive disposal cell. Officials responsible for remediating Weldon Spring contend that such public access generates trust in the remediation process and enhances public perceptions that the waste presents little risk of disturbing human activities. I have previously discussed this site (Bowers 2018), attending to the challenges of disclosing all of the actors involved in constituting the space. In this chapter, I situate Weldon Spring within the specific concerns of this book by drawing on notions of the sublime to describe how the Weldon Spring site and another site that offers a public encounter with acid mine drainage afford the opportunity to reorient public practices of environmental participation and responsibility. Spaces in which waste becomes publicly accessible afford the possibility to realize the limits of modern, institutional practices to control waste while also presenting the need to recognize the publics' role in surveilling and monitoring these heterotopia. While such public intervention characterizes sites of environmental injustice, the prevalence of waste and the emergence of publicly accessible sites that extend beyond geographically, marginalized locations reveal the need to reorient all publics to their contemporary, dirtied ecology and the relationality and responsibility that goes into constituting such sites. Recreational spaces of dirty play, as discussed in this chapter, afford the potential for a new ecological awareness in which humans better acknowledge their participation and responsibility as they dwell in dirty worlds. Shifting ecological awareness to the dirty and to the heterotopic also holds the possibility for publics to recognize the need for a more equitable encounter with waste. The resulting relational understanding provides the means to bridge the distance that has long defined and separated those facing environmental injustices since the distance between publics is no longer measured through the disparate subjects of producers and consumers but rather through subjects mutually implicated in dwelling in dirt.

In chapter 5, I discuss a final example of heterotopia and explore the potential and ethics of public agency in a dirtied world. More specifically, I illustrate the way that a happenstance configuration of objects that includes speedboats, sediment, weather events, and festival space congeal to offer agency to a public encountering toxic contamination. In this instance, agency emerges not through epistemological authority but rather through heterotopic, material arrangements and the rhetoric that the configuration enables. Due to changes in seasonal precipitation and the consistently high rates of sedimentation, the lake on which speedboat races are held has become too shallow for the boats to safely operate. Dredging, which will deepen the lake to accommodate the boats, will also, as residents contend, remove the toxic contaminants from a shuttered zinc smelter that rest within the lake bed. The publics' agency and resulting rhetoric are contested by an alternative stance, one also enabled by the unpredictable arrangement of objects. In this institutional rhetoric, the configuration of objects, especially the recurring and increasing rates of sedimentation, will address the toxins in the lake bed since deposited sediment will encase and safely contain the toxins from the defunct smelter. The series of sediment deposits will also transform the lake into a wetland and, following the frames familiar to environmentalism, will create an ecology beneficial to a wide range of human and nonhuman species. While the heterotopic configuration affords the possibility of these rhetorics and for publics to gain a sense of agency, ethical consideration must be given to the emergent rhetoric and realities that become constituted from the novel arrangement. Neither the rhetoric that advocates for dredging the lake or that which promotes the benefits of the lake's transformation to a wetland ecology will result in the removal of the toxins. Additionally, both efforts neglect the relationality inherent in the lake's dirtiness, as both dismiss the toxic contaminants that flow into the lake from the Illinois River, the lake's primary source of water. In sum, neither rhetoric results in the familiar clean and restored landscape that has traditionally defined the human relationship to waste. Similar to the discussion in the previous chapters, the exploration of the heterotopia in this chapter reveals the potential for novel forms of public participation and environmentalism within a dirty world while also advancing the need for a normative framework to understand the potentiality and constraints that novel, contemporary configurations afford humans and the ethics involved with respect to these emergent human responses. In this regard, this chapter extends my previous analysis (Bowers 2020) of this configuration by exploring the normative concerns that arise from a materially emergent rhetorical performance.

This book is not intended to serve as an epitaph for the familiar perceptions and rhetorics that have informed environmentalism and the publics' relation to waste. Efforts to generate an absence of waste, through forms of advocacy to limit emissions and other forms of detritus, prophesies of a despoiled ecology, and promises of an ecological attunement with the nonhuman to promote preservation, have provided meaningful intervention to restrain and even eliminate detrimental, human practices. And as climate change clearly illustrates, advocacy efforts to lessen and eliminate dirty emissions must remain a key practice for contemporary environmentalism. At the same time, however, we must acknowledge the inadequacies of these familiar environmental habits of thought, given their inability to provide for a world clean and absent of dirty matter. Concerns over the dominant influence of industry and their associated practices that contribute to a dirty world must surely continue, yet these conversations and attempts to revise imbalances of power and influence offer little to accommodate contemporary publics with the needed awareness and capacity to understand and live within a persistently dirty world. While publics may have been able to previously dismiss the existence of waste or move it to places unseen and populated by marginalized human and nonhuman populations, the ubiquity and obstinacy of the vibrant matter make this public stance of absence and removal no longer tenable. This material reality, therefore, invokes the need to move beyond the past perceptions, prophecies, and promises promoted by environmentalism and to instead find ways to dwell within this new ecology. This book attempts to provide some initial steps on that trajectory.

1

✢

The Inescapable Presence of Dirty Matter

On August 5, 2002, the Weldon Spring Site, located in the western suburbs of St. Louis, Missouri, became accessible to the public. Once a restricted area that housed various facilities to produce TNT and to process uranium for the US military arsenal, Weldon Spring has since been reconfigured into a recreational and cultural heritage space. Since 2002, the public has been allowed access to the remediated site, with opportunities to view various artifacts displayed in the Interpretive Center that document the site's contributions to the nation's defense and that also describe the efforts to eliminate the toxic contamination that resulted from manufacturing the weapons. Visitors to Weldon Spring, however, may encounter more than the cultural artifacts housed in the Interpretive Center; Weldon Spring also provides opportunities for the public to birdwatch, walk among native plant species, and hike and bike along trails that weave throughout the space. And at the center of these recreational activities is the disposal cell, a rock-covered mound where the legacy radioactive materials and other contaminants resulting from the production of the weapons are currently stored. Rather than transporting this toxic waste to a landfill distant from public view and access, officials at the Office of Legacy Management, a branch of the U.S. Department of Energy responsible for the site's remediation, decided to leave the contaminants on-site and to include the forty-five-acre disposal cell as part of the heritage and recreational experience. By way of concrete stairs and pebbled pathways, the public has the opportunity to climb atop a pile of radioactive and chemical waste and to hike and bike alongside its perimeter.

Climbing atop and hiking around a nuclear waste pile are not activities commonly associated with the way modern society has interacted with toxic substances. More familiar practices attempt to make toxic waste spatially distant from places of play and public life. Some have argued, however, that the absence of waste has fostered a particular modern, Western subject whose relationship with the discarded material reinforces environmentally destructive capitalistic and consumerist behaviors (De Coverly et al. 2008; Foote and Mazzolini 2012; Hawkins 2006; Rogers 2006; Royte 2005). Additionally, studies in environmental justice clearly demonstrate that the spatial practices that foster the absence of waste are rooted in class relations (Bullard 2018). Publics marginalized due to race and class find an absence from waste difficult to obtain given that the spaces they inhabit are more often contaminated by industrial waste and toxic materials. Such unequal distribution means that other members of the American publics "can more readily forget the costs of toxic pollution because they—the waste and the people disproportionately affected by the waste—appear hidden" (Pezzullo 2012, 121). But the continuation of detrimental industrial practices, the vibrancy of legacy waste, and the limited success of environmentalism have resulted in the geographic distribution of contamination and the resulting environs in which encounters with toxins are no longer restricted to specific classes.

Given the environmental and human harm that the absence of waste fosters, attention has turned to exploring the possibilities that emerge when waste becomes present to all publics. Cortez (2012), for example, calls for *"a mode of perception* that recognizes and affirms all the ways in which we have a continuing relationality to what we consume as well as discard, even as powerful economic and social forces work to obscure these lines of connection" (231). Hawkins (2006) also advances the need to emphasize encounters with waste, arguing that existing environmental rhetoric such as the apocalyptic has constrained the potential of making waste more present. For Hawkins, an object of waste embodies more than just destructive and horrifying potential; through a revised material and symbolic presence, waste becomes transformative.

> When waste is framed as dead objects and relegated to its proper place in the dump or garbage truck it often fails to provoke. It poses no questions to us because it has been regulated and rendered passive and out of sight. Waste as dead objects throws up few possibilities, but waste as things is full of promise, full of the possibilities of becoming a resource for being. (75)

Recirculation, according to Hawkins, offers one way that dirty matter may generate new relations to waste, evident in practices such as recycling and antiquing, where the material object of waste becomes present and reconstituted as it is (re)placed within new configurations. But similar to the limits of conventional environmental rhetoric, practices of recycling and reuse are unable to fully constitute contemporary public encounters with waste. Through its vibrancy and subsequent capacity to reshape the world, waste is becoming more present, pushing back against human intentions and challenging existing symbolic practices. Existing habits of representation and practices of reuse, therefore, do not fully account for the new configurations in which humans encounter waste and fail to address the resulting need for novel forms of perceiving our contemporary contaminated world.

Like Hawkins, I also draw from studies in "thing theory to investigate how waste's materiality might become present to us" (Hawkins 2006, 17). But rather than identifying the potential for waste to become rearticulated through its recirculation among new configurations, I emphasize the need to attend to waste's material and semantic constancy and its vibrancy as a detrimental actor who refuses to be influenced and reshaped even as it moves into new spaces and assemblages. In this regard, the increasing human encounters with waste allow for a consideration of the promises advanced by new materialism and the posthuman, particularly in terms of the emergence of new modes of perception and the resulting becoming of a more ethical human subject. In arguing for the capacity of configured material to influence human perception, Bennett (2010) refers to an instance when she observed a series of objects, including a dead rat, a thrown-away glove, and a plastic bottle cap on a Baltimore street. Upon seeing what many would perceive as individual, unrelated items of trash, Bennett instead "realized that the capacity of these bodies was not restricted to a passive 'intractability' but also included the ability to make things happen, to produce effects," made possible "because of the contingent tableau that they formed with each other, with the street, with the weather that morning, with me" (4–5). Novel perceptions of waste emerge not through the capacity of an individual, discarded object but rather by the collective force of the human and other-than-human objects that constitute the encounter. Bennett's attentiveness toward the particular configuration, for instance, emerges through her entanglement among all of the objects that constitute this particular space and time but also "a perceptual style open to the appearance of thing-power" (5). By Bennett's admission, the effects produced by the particular assemblage emerge not only through the particular configuration of human and nonhuman objects but also Bennett's perceptual proficiency that allows her to reimagine the configuration and in turn posit the possibility that such reimagining of matter can lead to an ethical, ecological form of new materialism. Subsequently, hopes that an encounter with waste may generate a new political subject must likewise take shape not only from the emergence of novel and influential spatiotemporal configurations but also through a subject's capacity to perceive such configurations anew.

This book explores the prospects for such transformation, focusing on the emergence of several contemporary, novel configurations in which waste is present and the means by which humans may perceive and act anew within such spaces. In the following pages of this chapter, I provide a brief survey of the perceptual apparatus available to American publics that currently informs their encounters with waste. My discussion includes how previous configurations of space and time afforded the opportunity for novel and more ethical perceptions and practices to emerge, configurations that prompted the familiar environmental practices and rhetorics that have guided the publics' understanding of and relation to waste. While the rhetoric, legislation, and advocacy efforts that resulted from these configurations have played central roles in fostering an awareness of the environmental dangers of industrial toxins, these common practices no longer afford publics the perceptual means to understand and act upon the reconfigurations and crises facing the contemporary world. I thus conclude this chapter by forecasting the potential of heterotopia as a conceptual starting point by which perceptions and resulting practices more fitting for the present era may be generated. A concept primarily traceable to Foucault, heterotopia are often characterized as other and alternate spaces. Hetherington (1997), for example, defines heterotopia as spaces that "organize a bit of the social world in a way different to that which surrounds them. That alternative ordering marks them out as Other and allows them to be seen as an example of an alternative way of doing things" (viii). The otherness that marks heterotopia in the posthuman age comes to be not just because of the presence of waste but also through the incongruity that emerges from the geographic, material, and disciplinary expansiveness that characterize contemporary configurations. A publicly accessible nuclear waste landfill that is also the site of recreational activities clearly presents an alternative way of ordering spaces of waste. But perceiving the site as a bordered, geographical space that mingles the two practices of play and disposal neglects the material displacements and dispersed entanglements that constitute the otherness and incongruity of contemporary heterotopia. Posthumanists argue that we can become more ethically and responsible subjects by recognizing the increasingly relational quality of being in the world; and, as I will sketch in the following chapters, the incongruous juxtapositions of distributed actors and praxes that constitute contemporary heterotopia offer an opportunity to explore the possibilities and constraints that such a relationally constituted world afford. But, as Bennett's encounter attests, such potential can only be realized by fostering the human potential to understand its capacity to responsibly act within such incongruous and novel relationality.

MODERN SOCIAL ORDER REQUIRES THE ELIMINATION OF DIRTY MATTER

A public's relation to industrial waste has been shaped not just by the danger to environmental and human health such material present but also by a range of historical practices and representations. While waste is often defined as matter with no value or use, it has also been aligned with other negative connotations, including terms such as soiled, rotten, and dirty. Dirt, as Douglas ([1966] 2002) explains in her heavily referenced anthropological study, is "essentially disorder" (2), since its material presence challenges a culture's dominant practices that aim to promote and maintain individual and thereby communal cleanliness. Attempts to constitute a public free from filth have consistently been informed by religious philosophies, which are guided by the central premise that "holiness and impurity are at opposite poles" (Douglas [1966] 2002, 9). From these religious principles emerge common symbolic and representational meanings; dirt becomes associated with acts of sinful behavior and cleanliness becomes coupled with conduct deemed spiritually proper and pious. While these symbolic meanings of dirt and purity serve as important resources to establish and maintain social order, the materiality of dirt also plays a significant role in fostering the proper citizenry. The oft-cited religious adage "cleanliness is next to godliness" may be interpreted symbolically, yet the efficacy of the maxim also derives from the material presence of dirt. "What appears to the onlooker's eyes to be the dirtiness of others generates clear moral and political judgments about their behavior and lifestyles" (Newell 2015, 40). The visibility of dirt on an individual's body socially constitutes and communicates an embodiment of immorality. Contrarily, a body cleansed so that the dirty matter is washed away and absent from a person's physicality constitutes a subject who exhibits the practices that reflect and maintain the proper civic morality of a culture.

In addition to the influence of religious tenets, secular institutions and practices have also played significant roles in promoting the relation between the absence of dirty matter from human bodies and the proper citizen. One of the more notable examples occurred during the initial entry of soap into the consumer marketplace. As Callahan, Lofton, and Seales (2010) assert, the most prominent means to promote soap's role in constituting the clean and noble citizen took place in the early twentieth century through The Cleanliness Institute, whose membership consisted of various soap manufacturers who used various educational and instructional messages to establish a "campaign on behalf of a commodity cast in the guise of moral sanitation" (22). The various publications produced by The Cleanliness Institute established rituals that instructed the citizenry in the proper use of soap, including the moral and civic value of its use. "The Cleanliness Institute's primary task was to train Americans in new cleansing rituals, to ensure that they not only knew about soap, but that they needed to use soap (in a prescribed ritual manner) in order to be good, godly Americans" (27). With the modern prominence of industrialism and scientific enlightenment, the authority to practice dirt avoidance and to generate the resulting proper and pure citizen no longer emanated strictly from religious tenets; cleanliness, in terms of one's physical appearance and even one's residence, became an idealized identity of the modern American citizen attainable through modern consumer goods and practices.

But as modern industrial practices provided consumers with access to an increasing volume of goods, including products that created the clean and proper citizenry, the waste that resulted from producing these consumer goods also grew, resulting in the need to manage and dispose of this most modern form of dirt. Addressing the dirt of modern industrial society has become, as Douglas ([1966] 2002) insists, an "effort to organize the environment" (2), rooted largely in practices that place the waste material so that it, like dirt on the human body, is visually absent. Hetherington (2004) offers a clear account of modern waste's spatial quality: "Disposal, I contend, is not primarily about waste but about placing. It is as much a spatial as temporal category. Terms like "waste disposal" and "waste management" are misnomers. Rather, disposal is about placing absences . . . and this has consequences for how we think about "social relations" (159). When members of a public no longer need or desire a certain consumer good or see it as soiled and dirty, they identify the commodity as waste and place the newly defined matter at the residential curbside. Similar to the dirty matter on the human body, industrial dirt is likewise removed and made invisible. But the cleanliness of one's environment is attained not through the mechanisms of some cleansing process or product but rather through the practice of emplacing the dirt elsewhere.

Such spatial practices of disposal constitute the moral citizen and disciplined consumer, subjects who appropriately discard refuse to maintain a clean and orderly household and community. And these practices of disposal also enable the human subject to enact the behaviors of the proper consumer. "The dynamic tendencies of consumer capitalism are particularly geared to the production of surplus and in order for the new to be accommodated, the old must be chucked out, erased or made invisible or else it will violate public and personal boundaries and propriety" (Edensor 2005, 315). These perceptions of waste and the performances of discarding the dirty, however, also mark the citizen as complicit in producing the harmful environmental impacts attributed to modern industrialism. "At an individual level, because trash is quickly removed to trash cans and collected from our places of residence (literally taken away from us) on a regular basis, this essentially relieves us from any further responsibility. This is what is referred to here as the *social avoidance of waste*" (De Coverly et al. 2008, 290). The modern propensity to materially and perceptually avoid waste extends beyond the human subjects' performances to remove valueless and dirty consumer items; distancing industrial waste from certain human, inhabited spaces also leads to the publics' lack of acknowledging the vibrancy of waste and the matter's potential to restructure reality.

The perceptions and practices that foster such absence and avoidance have been met with ample criticism. As a result, efforts have been made to reorient publics toward a social acknowledgment of and more responsible relation to waste. Recycling and its related practice of antique hunting have become popular practices to defamiliarize prevailing perceptions of waste. In these activities, consumer goods once deemed as outliving their value are resurrected by connecting the previously discarded product to some new use or some new commercial value. Central to these practices is the capacity to redefine the waste object, a practice, however, that may be informed by and subsequently reify class distinctions. For example, Hawkins (2006) discusses *The Gleaners and I*, a film that explores the practice of individuals combing through potato fields in northern France to collect the discarded vegetables deemed unfit to be transported to market. Through her interpretation of the film, Hawkins illustrates how waste is created and perceived through "the demands of the market, the restrictions supermarkets place on potato size and shape" (83). While the consumer market may generate a particular definition of an acceptable vegetable, those scouring the fields for the left-behind potatoes perceive the objects through an alternative economic perspective that affords the opportunity to re-assign value to the potatoes. In this example, matter becomes constituted by way of its circulation within alternative economic practices. What is waste depends upon economic class.

Recycling and reuse, while perhaps lessening the volume of material defined as waste, do not take into account the vibrancy of dirty matter and its capacity to resist semantic and representational alterity. Certain objects classified as waste may indeed become transformed as they circulate within new assemblages; yet objects deemed as waste can also enter new configurations and still maintain their existing dirtiness and dangerous influence. For example, some forms of industrial waste may become recycled and regain commercial value, but redefining the dirty matter of industrial toxins seems constrained not just because of waste's material properties but also because human relations to dirty matter are affected, quite appropriately, by horror and disgust. Schneider (2012), for instance, discusses the city of Milwaukee's efforts to remediate and market the city's sanitation and sludge as fertilizer, noting that "marketing products from sewer sludge, however, continued to face deep-seated antipathy from users toward including human waste in products for human consumption" (178). If industrial waste is resistant to redefinitions and its ubiquity presents challenges to any persistent attempts to keep all bodies clean and absent from the matter, then alternative perceptions and practices must be developed to equip society with some means to reconcile the appearance and presence of such dirt.

FROM DIRTY CONFIGURATIONS COME NEW PRACTICES

The emergence of the practices of risk management and environmental advocacy can be traced to unregulated industrialism's march through the twentieth century and a series of altered spaces that occurred as a result of waste's increasing presence. For instance, the configuration that emerged in 1948, in Denora, Pennsylvania, a town situated within the heavily industrialized corridor along the Monongahela River south of Pittsburgh, Pennsylvania, is often seen as playing a central role in initiating practices to alleviate the consequences for health of air pollution caused by industrial emissions. While air pollution had been visibly present during the late nineteenth century and increasingly so into the mid twentieth century, largely due to the nation's reliance on coal, the industrial dirt emitted into the air had little influence in fostering any significant political response. It was only through specific configurations that generated the influential presence of waste and the subsequent public perception that disturbed the prevailing social order. In the case of Denora, for example, the entanglement of specific atmospheric conditions and the unique geography of the region trapped emissions from the surrounding steel and zinc industrial facilities within the Monongahela River Valley, resulting in a layer of toxic smog that hung over the town.

Lasting for four days in October, the lethal fog was deemed responsible for the deaths of nineteen citizens, with some suggesting the death toll would have been higher had not rain dissipated the deadly smog (Boissoneault, 2018). The increased presence of dirty air during the mid twentieth century eventually led to a series of legislative acts in the United States designed to curtail the volume of deadly industrial emissions. But, as many have suggested, it was not until the passage of the Clean Air Act (CAA) in 1970 that significant action took root to address the impacts of industrial waste in the air (Mosley 2014, 155). New practices emerged to guide the CAA, notably in the form of scientific processes to determine the amount of the waste particles that pose a threat to human health and to measure the amount of these emissions that existed in the air. As long as the quantities of dirt in the air fell below the identified standards and acceptable levels of emissions, the emitted waste was perceived to present no detrimental effects on humans. Dirtied air existed, but its limited presence was deemed as nothing to worry about. Science thus became an instrumental actor in environmental policy and in the publics' perception of waste, leading to current practices such as risk assessment, risk communication, and risk management. In addition to the practices used to determine the extent of risk posed by certain quantities of pollutants and waste, the development and use of objects also became crucial as technological innovation allowed industry to maintain desired levels of production while also meeting the required pollution standards and levels of emissions. Certain configurations in which waste becomes present thus can be seen as agents of change, fostering a range of new perceptions and practices. Borrowing terminology from Barad (2007), waste is thus granted a material dynamism, "generative not merely in the sense of bringing new things into the world but in the sense of bringing forth new worlds, of engaging in an ongoing reconfiguring of the world" (170). Through the disruptive political incursions that emerged through these configurations of waste, new worlds and spaces came into being as specific legislative actions, habits of representation, and advocacy efforts lessened the amount of sulfur dioxide from coal emissions, eliminated lead gasoline used in cars, and banned particulates and chemicals that were found to deplete the ozone.

The emergence of new perceptions and practices can be traced to more than just those configurations and spaces of air pollution. New practices emerged in relation to the presence of industrial waste in the nation's waters and in the various spaces contaminated from the legacy of the unregulated dumping of industrial toxins. For example, in the 1960s, the interaction of waste with the nation's various streams, rivers, and lake waters brought the issue of water pollution to the public consciousness. One of the most influential disturbances occurred through the configuration that generated foam and soapsuds in various waters, a product of the "skyrocketing sales of laundry detergent" fostered by the "power of advertising to differentiate products ('cleaner than clean') that had been reshaped by new technologies" (Gottlieb 2005, 118). The entanglement of industrial production, marketing, and consumer behavior played a key role in configuring a new world in which the nation's waters took on an alternative form. In an ironic twist, the desire for cleanliness, that long-held order that is so central to maintaining proper societal bodies, helped generate this particular presence of modern, industrial dirt. The presence of waste also became evident in other instances, visible in the eutrophication of many waterways, especially the Great Lakes, and in the flames that emerged atop of the Cuyahoga River. As Gottlieb describes: "By the 1960's water contamination problems were being magnified by such sources as untreated sewage and industrial wastes. These flowed freely into such places as Lake Erie and smaller lakes and streams, polluting them to the point of eutrophication" (118). The entanglements that generated the dirtied water eventually created enough disturbance to lead to specific legislation in the shape of the Clean Water Act (CWA). Like its partner the Clean Air Act, the CWA, along with the discourse and performance of environmental groups, also had some initial success in moving the public, industry, and government toward a revised sense of environmental responsibility, with the prohibition on phosphorous and the resulting restoration of the Great Lakes a testament to the impact of the regulations and new bodies and practices.

While the CWA and CAA emerged by way of the presence of waste in the air and water, additional new perceptions and practices arose in the middle part of the 1970s due to various configurations involving buried toxic waste. While industries were emitting toxic by-products into the air and water, other toxins that had been created by retired manufacturing facilities and made absent through unregulated disposal practices publicly emerged, with Love Canal serving as a landmark case. The planned, idealized neighborhood in Niagara Falls, New York, named after its developer William Love, failed to fully materialize due to a misperception that the geography of the region would generate suburban prosperity. In discussing William Love's vision of the planned residential enclave, Beck (1979) writes that "he [Love] felt that by digging a short canal between the upper and lower Niagara Rivers, power could be generated cheaply to fuel the industry and homes of his would-be model city" (17). But as the potential of hydroelectric power was diminished due to advances in other forms of generating electricity, the canal was abandoned and converted into a landfill for municipal and chemical waste. In 1953, Hooker Chemical, which owned the landfill and was using the canal to dispose of toxic waste generated during chemical production, sold the property to the city for one dollar. City officials subsequently developed the property, covering the canal and chemicals that rested in the crevice with fill material, and then built 100 houses and a school atop the buried toxic waste. The residential development that came into being atop and surrounding the landfill took the guise of mid-twentieth-century American suburbia: "Perhaps it wasn't William T. Love's model city, but it was a solid, working-class community" (Beck 1979, 17). As Lois Gibbs ([1982] 2011), Love Canal's most famous citizen and environmental advocate, acknowledges: "When I first moved to the Love Canal neighborhood in Niagara Falls, New York, the only thing it symbolized was 1970s suburbia. With lots of green tress and children playing outside, I thought it would be a peaceful, safe place to raise my family." (1) Whether in Love's planned vision of a utopian city or in Gibbs's initial impressions of the suburban Eden, these perceptions of the space of Love Canal were rooted in the seeming absence of dirt and waste.

Yet this image of suburban purity and cleanliness was disrupted through the movement and subsequent presence of the buried waste. While interring industrial waste was deemed as an appropriate disposal method at the time, the industrial toxins eventually leached from the buried drums and found their way into the surrounding neighborhood yards and water. As described by a visiting EPA agent, the place as formed by the toxins clearly stands opposed to any vision of purity:

> I visited the canal area at that time. Corroding waste-disposal drums could be seen breaking up through the grounds of backyards. Trees and gardens were turning black and dying. One entire swimming pool had been popped up from its foundation, afloat now on a small sea of chemicals. Puddles of noxious substances were pointed out to me by the residents. Some of these puddles were in their yards, some were in their basements, others yet were on the school grounds. Everywhere the air had a faint, choking smell. (Beck 1979, 17)

The movement of the chemicals, a trajectory unanticipated and uncontrolled by human actions, not only generated a new space, one now constituted as dirtied and soiled, but also reconfigured human bodies, reconstituted in the form of cancers and birth defects.

Though it took a concerted effort on the part of the citizens of Love Canal, actions were taken to address the disturbing and horrific reality that the toxins had brought. Residents of Love Canal were displaced and distanced away from any continual encounter with the contamination. The Comprehensive Emergency Response, Compensation, and Liability Act (CERCLA), more commonly known as Superfund, came into being in response to the acknowledgment of numerous spaces spoiled by buried and discarded toxic waste. Additional legislation and subsequent practices emerged in response to the increasing volume of toxic waste being produced and becoming increasingly present. While legislation such as the CWA and CAA attempted to regulate and address waste in the air and water, "the number of new and possibly toxic chemicals entering the market each year was creating enormous stresses on this regulatory system, given its inability to manage the range of hazards presented by these new chemicals" (Gottlieb 2005, 244). The passage of two pieces of legislation, the Toxic Substances Control Act (TSCA) and the Resource Conservation and Recovery Act (RCRA) laid the groundwork by which "the triad of mainstream environmental group, government regulator, and pollution control or waste management industry promoted these hazardous waste management laws and regulations as an effective way to deal with the growing problem of toxics in society" (245). The 1970s can thus be seen as a compelling arrangement of time and space in which industrial waste became an influential public actor, spurring the generation of various novel perceptions and practices that subsequently informed the way modern humans would dwell with dirty matter.

While the perceptions and practices that emerged from these various reconfigurations of space became foundational to understanding and addressing the dangers of industrial dirt, they nonetheless, eventually served to reinstate the belief that the prevailing order could again be restored by making waste and its affiliated harm and disgust absent from public encounters. The legislation along with the corresponding practices that emerged were all aligned to sponsor the belief in the human capacity to purify a landscape tarnished and spoiled by the waste, whether that landscape be in the form of the air, water, or places in which a public lived, worked, and played. If the waste could not be removed and shipped elsewhere, then the humans who may encounter the dangerous matter would be removed from the toxic configuration. If those actions were not taken, then the scientific processes central to risk assessment could be used to determine that the amount of the contaminants was below the standard identified as being detrimental to biological processes, in effect, making the waste absent by its incapacity to do bodily harm.

But given the current capacity of waste to encroach upon and disturb all walks of contemporary life, modernity and its associated practices (to include environmental advocacy) to address dirty matter have to be deemed as insufficient and inappropriate to the current epoch. For instance, in the posthuman reality, an increasing number of summer days, and even more recently winter days, are defined as air advisory days because the pollutants and dirt in the air can pose harm to various bodies. During such designations, publics are urged to refrain from certain activities, and those with respiratory distress are especially cautioned to take heed of the dirty air. These air quality days result not just from car emissions but also emissions from industrial facilities and coal-fired power plants and the geography and meteorology that traps and helps to congeal the emitted particles. Yet such contemporary "bad-air" days do not have the same influence as those that occurred in Denora, Pennsylvania, or elsewhere during the mid twentieth century. Similarly, while Lake Erie has in many ways been revitalized and resurrected from the eutrophic state that characterized its being in the 1970s, it is presently being threatened, as are numerous other national waters, by another presence of dirty matter that remakes the water into a eutrophied body. Phosphorus again, not from the use of detergents but rather through the use of fertilizer, has marched its way in significant volumes into lakes and other waters, with swaths of dead zones created by the assemblage of the chemicals, algae, precipitation events, and warm water temperatures.

And while the Superfund Act has attempted to clean up various polluted landscapes, scores of dirtied spaces remain on the list waiting to be remediated while other sites are yet to be identified as the waste funneled and stored away from retired industries lies in wait for a future public incursion. As Brown (2019) contends, "Every territory on Earth is the site of some former ecosystem and of contaminants produced in the Anthropocene" (S198). While humans have exerted considerable intellectual energy and economic resources toward making waste absent, the dirty and dangerous matter increasingly no longer willingly accepts human directives. And as the continuation of crises brought about by waste's presence attests, the promises of posthumanism and the emergence of new environmental subjects and practices have also failed to materialize. Posthuman spaces reconfigured by material forces such as waste have not generated the kind of perceptual and political change that occurred during the late 1960s and mid 1970s. As I will contend, such failure can be attributed to the reliance on familiar figurations and practices of environmentalism and their resulting perception of ecological care and responsibility. Attention must therefore be directed at acknowledging the irrelevance of prevailing forms of environmentalism and toward identifying more appropriate and meaningful ways that environmentalism can contribute to dwelling within the reconfigured, contaminated world.

WASTE'S FORAY INTO THE POSTHUMAN LANDSCAPE

Because of its undeniable transformative potential, toxic waste serves as an illustrative example of the agency of nonhuman matter; but the continuing incursion of dirty matter into public life and the continual remaking of spaces illustrate that the potency of the material's influence to generate environmentally ethical subjects may be overstated or misunderstood. If new configurations, especially those in which waste is a primary actor, have the potential to promote change toward a good, we also need to recognize, however, that matter's influence may be constrained within various spatiotemporal assemblages by newly, entangled relations but also because of the continued reliance on familiar perceptual frames that attempt to order dissonance into accustomed understandings. Inquiry, therefore, must be directed at the way that matter's influence in shaping configurations may move perceptions away from the common ethical considerations aligned with environmentalism to novel, normative positions. This book thus acknowledges the role of matter in shaping the world but also, through the case studies explored in the coming chapters, affirms that fostering novel human perceptions and actions remain central to the realization of any alterity made possible by new configurations. The promise of new materialism is clearly stated by Coole and Frost (2010):

> It is evident from new materialist writing that forces, energies, and intensities (rather than substances) and complex, even random, processes (rather than simple, predictable states) have become the new currency. Given the influence of classical science on the foundations of modern political thought, it is germane for new materialists to ask how these new conceptions of matter might reconfigure our models of society and the political. (13)

Granting such transformative potential to these novel configurations raises two points of contention, however. First, admitting that other-than-human objects may shape the human subject could suggest that the evolving human body is determined by the force of the configurations and material objects; the potential for a more ecologically aware human subject, as a result, is constituted not through intention or conscious human action but rather the subject's opportune position within a fortunate configuration. As my brief historical survey of environmental advocacy illustrates, emerging configurations have influenced the human subject to act in more ecologically responsible ways. The contemporary configurations in which waste is quite present would seem to also carry some opportune level of influence and agency, yet these spaces have not been able to invoke similar, caring subjects and practices generated decades ago. If agency is granted to material assemblages then the different responses noted above suggest that either ethical action is provoked only through the right configurations or that human actors, much like their material conspirators, have the potential to push back against the influence of material others. An appropriate line of inquiry thus concerns the manner in which the human agent operates within these configurations and within assemblages of varying influence. And such a line of inquiry is especially pertinent to any investigation of novel practices and the emergence of any alternative ethic of ecological care.

Therefore, we must resist assumptions to grant any movement toward ethical and responsible care for the planet with happenstance and the right spatiotemporal configuration. The environmental disturbances that generated the various legislative and advocacy efforts during the 1970s and the inaction that characterizes contemporary incursions of dirt cannot simply be distinguished by way of the more potent influence of those earlier configurations. The prominence given to practices and perceptions that favor industrial production and consumer behaviors clearly have some influence on constraining contemporary concern over dirty matter. Yet the familiar habits of perception engendered by the increasingly inappropriate environmental rhetorics that took root during those initial configurations also share a considerable amount of blame in limiting the capacity of the contemporary presence of dirty matter to invoke concerned subjects. As criticism of the existing consumerist and industrial practices continues to fall short in generating more responsible practices, we are left with identifying new figures and rhetorics that better locate the human within these posthuman, dirty configurations and that in turn promote the needed perceptual frames by which to better understand the prospects of the human within such entanglements. Humans cannot wait for a material deliverance of the ethical, nor is such a prospect illustrative of the making of the world. New bodies and practices result from the commingling of human and material creativity and the proficiency of the human to perceive the novel possibilities the world affords. The fact that humans must dwell with dirt as opposed to simply making the matter absent or hoping for a return to some purified Eden calls upon the human to locate creative opportunities to exist within its contemporary, dirtied world. Moving such habits of perception begins with reorienting humans to waste as vibrant matter.

Drawing from Robert Sullivan's description of a New Jersey garbage pile, Bennett (2010) directs attention to the vibrancy and continued presence of waste, even as the matter finds itself within human practices of absence and disposal. "Sullivan reminds us that a vital materiality can never really be thrown 'away,' for it constitutes its activities even as a discarded or unwanted commodity" (6). Adhering to the practices of absence demands a dismissal of matter's agency and advances the primacy of humans to control, contain, and erode waste's capacity. Yet, as new materialism suggests, waste can provoke even when it is absent. Granting a presence to waste equates, however, to acceding that the material we refer to as industrial discharge is not innate matter waiting to be transported to a specific place nor can it be readily ordered by placing it within some human-designed container. Nor can dirty matter always be readily recycled and given alternative value. Waste has the capacity to resist and even express itself in ways beyond human control. Moreover, its capacity to act and invoke new realities cannot be adequately assessed by way of scientific assessments of waste's volume and quantity; its agency and presence must instead be recognized as emergent through the relations in which it becomes entangled. The key contributions that new materialism adds to any recognition of waste's agency and waste's capacity to reshape human subjects rest in the assertion that an object's ability to act may extend beyond any human constraints and is a product of relations as opposed to a capacity inherent in an individual object. Moreover, waste, as vibrant matter, may limit human action, constraining the familiar and, as a result, redirecting perceptions and resulting normative concerns.

Therefore, the novel becoming of the world occurs relationally, through what Barad (2007) refers to as intra-action.

> The notion of intra-action recognizes that distinct agencies do not precede, but rather emerge through, their intra-action. It is important to note that the "distinct" agencies are only distinct in a relational, not an absolute sense, that is, *agencies are only distinct in relation to their mutual entanglement; they don't exist as individual elements* [emphasis in original]. (33)

Bergthaller (2014) further clarifies the centrality of the relational to new materialism and matter's agency, emphasizing that "new materialisms proliferate connections, point out that entities that seemed to be self-contained are in fact enmeshed in a tangle of relations to other entities, and demonstrate that they acquire their seemingly intrinsic qualities only within these relations (41). When emplaced within configurations and entanglements with novel and seemingly incongruous objects, waste may acquire significant influence in generating new spaces and, in turn, new worldly consciousness which may not readily translate into familiar ethical frames. Such a chance for alterity and creativity coincides with the needed concession that these configurations of opportunity may emerge in an unpredictable manner, through actions and effects not directly assigned to the human agent.

> Many causal relations are not susceptible to either efficient or mechanical modes of analysis. There are efficient causes, as when, to take a classic example, one billiard ball moves another in a specific direction. But *emergent causality* [emphasis in original]—the dicey process by which new entities and processes periodically surge into being—is irreducible to efficient causality. It is a mode in which new forces can trigger novel patterns of self-organization in a thing, species, system or being, sometimes allowing something new to emerge from the swirl back and forth between them: a new species, state of the universe, weather system, ecological balance, or political formation. (Connolly 2010, 179–180)

The political capacity of matter thus emerges as an unexpected and novel configuration of time and space, calling upon human actors, especially, to locate the perceptual apparatus needed to respond to and act within such strange space. As I again want to stress, new politics and subjects do not marvelously emerge out of such accidental configurations; rather, alterity takes shape only when subjects develop the perceptual proficiency to recognize and respond to the novel possibilities afforded by these entanglements.

The political becoming made possible by reorganized material can be clarified by Bennett's (2010) new materialist reading of Ranciere. As Bennett describes, the disturbance fundamental to a political event can be understood as an "interjection by formerly ignored bodies" (105) that generates effects: "a political act not only disrupts, it disrupts in such a way as to change radically what people can 'see': it repartitions the sensible; it overthrows the regime of the perceptible" (106–107). The political interjection and disruption to the prevailing order occur when subjects previously absent from the public sphere, which Ranciere limits to the human, become present to the extent that they constitute an influential political body. The incursion of these previously absent political subjects, which Bennett, however, contends may include the nonhuman, make visible a societal problem by jarring common modes of perception. The new configuration brought about by the presence of the previously absent actors calls upon people to see the world anew. As I have noted, the presence of waste during the 1970s can be seen as prompting not just new worlds but also new perceptions and subsequent political actions. While environmentalism and its associated ethic of care for the natural world emerged from this new reality, so too did other practices, including the reliance on techno-scientific innovations to lessen the amount of industrial waste circulating through the environment and the prominence of scientific knowledge to assess and quantify the dangers of toxins.

But neither environmentalism nor techno-science have prevented the world from its current contaminated condition. The amount of dirt in the air, water, and land may have been lessened by these emergent practices, yet the continuing and ever-encroaching presence of the dirty matter in contemporary spaces reveals that the modes of perception that these practices foster are no longer sufficient to help people dwell with the ubiquitous presence of dirt. These resulting technologies and procedures afforded society the ability to perceive the world as clean and absent, at least conceptually, from dirty matter. Further, these practices have enabled a perception that human ingenuity and innovation can restore the previous pure and natural order of the world, a mode of thought commonly ascribed to contaminated sites that have become restored and remediated. Yet, as my previous discussion illustrates, even with such hope and promise invested in technology, numerous landscapes, waterways, and the air have been and remain contaminated and far removed from the vision of a clean and purified Eden. Environmentalism has fostered other modes of perception to order human relations to waste, notably in the symbolic prophecies of the coming dirtied and uninhabitable Earth. This habit of thought is commonly invoked by apocalyptic rhetoric, which advances the understanding that the presence of industrial contaminants will lead not to any sense of ecological equilibrium but rather to ecological disaster. Discourses of the apocalyptic, especially that of Rachel Carson, who many reference as the origin of the practice, have had a significant impact on raising attention to the dangers of industrial toxins and have also helped lead to the elimination of certain chemicals from industrial production. But the discourse has failed to maintain such a level of influence, owing largely to the failed prophesies that have been central to the rhetoric. Like the legislative actions that took hold in the 1970s and the promises of technology, the apocalyptic no longer offers the means to fully understand one's emplacement within the increasingly contaminated contemporary world. Given the ubiquity of dirty matter, humans find themselves in a novel place of incongruities and in need of a new mode of perception.

Subsequently, the emergence of new perceptions is an acknowledgment of the other. Waste can no longer be viewed as absent but rather as vibrant materiality, present in the everyday activities of all individuals. Moreover, its acknowledged presence need not be restricted to the common and familiar desires of cleanliness and the urge to rid one's environment of the dirty matter. Waste has shown and has made clear that it does not want to go away. Therefore, humanity faces the challenge to exist with this other and to acknowledge its capacities to influence, prompt, and constrain human actions. An acceptance of the dirty other does not equate to acquiescing to the detrimental policies that generate its abundancy. While I staunchly support efforts to lessen and even eliminate the production of toxins and waste, their removal from our everyday lives is unlikely given the prevailing influence of the practices that sponsor their production and the known and unknown legacy materials that currently circulate around and among us. We can continue to strive for a cleaner world, yet we must also accept and attend to the existence of the dirtied spaces we all inhabit and visit and find some means to ethically dwell within these contaminated spaces. Such prospects need not lead us toward despair, however. Drawing on posthumanism, Braidotti (2013), for instance, looks to the possibilities that exist as worldly material comes to be more fully present, possibilities that challenge the "narrow and negative social imaginary" (64) of the apocalyptic. Central to Braidotti's optimism is the formation of a posthuman subject, a body that is "geo-centered" (83) and not a product of isolated reason. "We need to visualize the subject as a transversal entity encompassing the human, our genetic neighbors the animals and the earth as a whole, and to do so within an understandable language" (82). Central to any new perception is the acknowledgment of a relational position of being present with human and nonhuman others within a contaminated world and a corresponding responsive to waste and various other human bodies and nonhuman vibrant matter. Given the unpredictable circulation of matter and the extent of the geographically dispersed relations and entanglements that human bodies and nonhuman matter now find themselves within, we are all likely to find ourselves engaging with a surprising and perhaps immediately dissimilar myriad of dirtied others. Latour (2011) lays the path to help recognize that our contemporary lives are constituted by dispersed and supposed incongruous relations with others:

> Everyday in our newspapers we read about more entanglements of all those things that were once imagined to be separable—science, morality, religion, law, technology, finance, and politics. But these things are tangled up together everywhere: in the Intergovernmental Panel on Climate Change, in the space shuttle, and in the Fukushima nuclear power plant. (21)

And industrial waste and toxins have become central actors within these diversely populated configurations. Perception within a contaminated world then becomes tied to human creativity and the capacity to reorient and act anew among the unexpected, seemingly dissonant entanglements with previously unseen, unknown, and separate others who refuse to disperse.

The occurrence of contemporary entanglements marked by juxtapositions with previously invisible and dissimilar others situates the posthuman subject within heterotopia. It is not just the assemblage of heterogeneous embodied and material others within bordered and confined space that marks contemporary entanglements and heterotopia. Contemporary heterotopia are also constituted by the mingling of discordant sites and practices that may be geographically distributed but are, nonetheless, relational. Foucault ([1967] 1984), for instance, notes the sacredness of the boundaries of such spaces of work and leisure and the private and the public (23), yet these once distinct spaces and the practices inherent to these spaces have since become entangled and increasingly less separable. The reconfiguration of space and more notably the juxtaposition of what were once separate and other objects, spaces, and practices present the hermeneutic challenge and opportunity that can spur novel perceptions to better equip publics as they encounter contaminated spaces. As recent efforts to align environmental rhetoric and environmental care with new materialism attest, change should no longer be understood as a result of actions by a single agent or even an assemblage of isolated human actors or human institutions; instead, the possibilities of new practices and bodies emerge through the relations within novel assemblages of dispersed human and nonhuman bodies and practices. "We cannot understand change if we examine the causal impacts of pre-constituted bodies, subjects, and environmental—rather we must attune to the circulation of affective, relational, or ambient forces that constantly articulate and rearticulate these provisional entities" (Wells et al. 2018, 21). But here as well, as the language used by Wells et al. concedes, such change is only possible through human "attunement." Generating a relation to and a responsiveness to the other and changes in praxis that may result from such attachments remain a human-driven endeavor. The emergence of practices that sponsor greater ecological care and concern for the other remains embedded within human perception which can then inform the human capacity to generate appropriate ways of living within spaces of incongruity.

Emphasizing human creativity and perception affords the opportunity to engage in one of the major criticisms aligned with new materialism. Critics of new materialism have questioned the way human interaction with nonhuman objects actually fosters a new politics and the possibility that novel configurations are prone to inherently invoke more just and ethical bodies. For example, Washick et al. (2015) contend: "That the new materialisms' ontology entails relationality is clear, but why relationality dictates a particular ethical position is not" (73). Rekret (2016) similarly challenges the ethical promise associated with new materialism, noting that the "ontological primacy of matter is grounded upon the call for a spiritual or ascetic transformation" (227) with such transformation occurring through "the subject's cultivation of the right affects" (229). Rekret does not discount the vibrancy of matter nor the potential of configured objects to invoke change; rather, he questions the mystical influence granted to vibrant matter that consistently moves human subjects toward a greater good. Once the perceptual interference is cleared and the human subject can attune properly to matter then the right and ethical affects are generated. For new materialists, matter seemingly possesses little or no capacity to move people toward lesser affects or to invoke questionable norms. Bergthaller (2014) echoes the concerns about assigning ethics to matter's agency, calling attention specifically to Bennett's claims about matter's normative superiority:

> The ethical implications of the material turn are much less clear and a good deal more unsettling than her upbeat rhetoric sometimes suggests. The realization that matter has agency offers no more ethical guidance than the attribution of intrinsic value to all living things—it merely begs the question how exactly, then, human value and human agency are to be weighed on the onto-ethical scales. (38–39)

Granting configured objects the potential to disrupt existing practices and to generate new orders does not dismiss the vitality of human actors; instead, the agency of matter calls for a reconsideration of the potential courses of action of human actors and the way humans perceive and act upon the possibilities that emerge through reconfigurations.

Sustaining the human agent's influence in these configurations does not equate to restoring the modern tendency to privilege human agency, however. The potency of an agent's influence remains a central concern of any exploration of material reconfigurations, yet such an investigation acknowledges that power and influence are not only materially and bodily dispersed but also manifest through one's capacity to interpret and act upon new and potentially highly influential possibilities that emerge in novel arrangements. Acknowledging the distributed nature of influence and the value of a hermeneutic proficiency in relation to one's entanglements addresses the challenge of some who suggest that new materialism neglects the way that power is unequally dispersed in the world.

> We would agree with Latour's analysis of the embedded situation of humanity, yet as with Bennett's work, the effect of a network analysis is to flatten out relations between different actors and actants. This has the impact of obscuring power relations between different actors, and in particular the possibility for one actor to alter the context in which others act. (Cudworth & Hobden 2018, p. 15)

Bennett (2010), however, clearly acknowledges the role of power in her discussion of vibrancy and distributed agency. Assemblages "have uneven topographies, because some of the points at which the various affects and bodies cross paths are more heavily trafficked than others, and so power is not distributed equally across its surface" (24). Bennett adds:

> Of course, to acknowledge nonhuman materialities as participants in a political ecology is not to claim that everything is always a participant, or that all participants are alike. Persons, worms, leaves, bacteria, metals, and hurricanes have different types and degrees of power, just as different persons have different types and degrees of power, different worms have different degrees of power, and so on, depending on the time, place, composition, and density of the formation. (109)

Even in attempting to maintain an emphasis on nonhuman matter, Bennett asserts that the influence of humans may vary within different configurations. "Humans and their intentions participate, but they are not the sole or always the most profound actant in the assemblage" (37). And as I argue, the agency and level of influence of humans that emerge from any reconfiguration may be influenced by nonhuman members of the assemblage but also by other humans that also populate the ensemble. For the configurations I explore in this book are not only constituted by a range of nonhuman matter, including waste; these heterotopia are also comprised of a heterogeneous and seemingly incompatible collection of human actors and their practices. Understanding agency and any emergence of novel practices pertaining to ecological care thus entails a consideration of the range of nonhuman matter and the range of human actors that collectively encounter, perceive, and act upon the resulting capacities that emerge within the contemporary crises posed by industrial waste.

In the remaining chapters of this book, I wish to take up the transformative potential of heterotopia within the posthuman landscape and to also offer an initial effort to respond to Braidotti's (2013) call: "The collapse of the nature-culture divide requires that we need to devise a new vocabulary, with new figurations to refer to the elements of our posthuman embodied and embedded subjectivity" (82). This book offers heterotopia as an appropriate figure to investigate and understand the contemporary human subject's embeddedness within a contaminated world. As noted, perceptions of the human relation to waste remain informed by notions of modernity, either in terms of its associations with human innovation as a means to maintain ecological balance and health or by the apocalyptic with its portents of ecological catastrophe; such perceptual stances are incompatible with a world constituted by dissonance and contamination. Additionally, familiar perceptions of space and their associations with environmental care are similarly unsuitable to understand the current configurations humans encounter. The images of the blue planet made available through space exploration of the 1960s and 1970s have helped shape an environmental ethos by reconstituting our notions of place as global and interconnected. Moreover, investigations of globalization have had additional influence in shaping a relational view of space as local practices become relationally associated with activities occurring elsewhere. This relational view of space and corresponding ethic are perhaps most evident in matters of environmental justice where the local consumption of industrial goods is cast within spaces of global production in an effort to disclose the seemingly distant and absent environmental impacts of industrial production and commodification. Such a view of space and the potential to make present environmental harm often use the figures of the networks of production and consumption (Heise 2008, p. 65). But contemporary configurations of waste require an additional view of space, an alternative figuration whereby the network or assemblage is not defined merely in terms of consumers and producers; instead, contemporary spaces must be defined as those entanglements in which various, heterogeneous matter (both human and nonhuman) share crises and find themselves entangled within collective spaces of contamination and incongruity. Space as figure and as interpretative frame thus entails a consideration of how these various actors perceive and act within these shared crises and the way that these perceptions and actions constitute contemporary, eco-ethical subjects emplaced and entangled within a dirtied world.

2

The Potential of Dissonance and Heterotopia

I spent my early adolescence in Spring Valley, Illinois, a rural town in the north-central part of the state. Spring Valley was bordered by the Illinois River to the south, level farmland to the west and the north, and a meandering growth of woodlands that ran alongside Spring Creek on the village's eastern end. For preteen boys during the 1970s, Spring Creek, with its pools of water shaded and cooled by the canopy of trees that lined its banks, provided an escape from the hot and humid Midwestern summers and a space removed from the adult rules that structured youthful behavior within the city's grid of residential streets. But within this wooded landscape on the eastern edge of town, Spring Creek was not the only location for unsupervised acts of recreation. At the eastern end of the village's main street, just beyond the last of the middle-class houses that lined the road, stood a roughly seventy-five-foot-high mound consisting of thin, irregularly shaped fragments of pink shale, a slope everyone in town affectionately referred to as "the dump." The pink-hued shale fragments were deposited not by glaciers or some other natural process but rather by miners clearing what was known as Spring Valley Coal Mine #1. During the late nineteenth and early twentieth centuries, coal mining became a prominent industry in north-central Illinois. While lesser in quality and amount than the coal mined in southern Illinois, the coal seams in the northern part of the state became valuable based on the seams' proximity to the growing metropolitan and industrial areas of Peoria, Moline, and Chicago (Kernc, 2011). To access the underground coal seams, miners first removed the shale that lay atop the coal, piling the pink matter into mounds away from the mine. Once the thin coal seams in the region were depleted and the mining industry left these rural towns, the hills of unusable refuse remained and became the primary, visible reminders of the former industry. And like Spring Valley, many other villages throughout the north-central part of Illinois also found themselves with similar pink landmarks.

For adolescent boys in Spring Valley, the dump was more than just a pile of unusable rock and leftover waste from an abandoned industrial practice. While the industrial refuse was located beyond the residential borders of town, a placement that attempted to make the waste absent, the pile of pink shale was quite present in many of our day-to-day youthful activities. Due to the particular slope of the hillside, the force of gravity, the crevices that were etched into the dump from years of rain and snow, and youthful imaginations, the dump was transformed from a pile of indifferent waste to a space of play and wonder. The reconfiguration of the slag pile in Spring Valley was not unique, however, as similar pink-shaded dumps located in neighboring cities also became more than just waste piles on the edge of town. For example, in Toluca, Illinois, a small village approximately forty miles south of Spring Valley, residents protested the removal of what they referred to as "jumbos," contending that the refuse pile conveyed a vital part of the community's identity. The *Chicago Tribune* documented the efforts of residents of various towns where jumbos were emplaced, noting how the encounters with the slag piles contributed to performances specific to a community: "As kids we used to climb ours all the time," recalled Roanoke Mayor Joe Amigoni. "When we got older, we cut a road up it with a 'dozer and sometimes we'd take old cars up to the top and push them off. It was quite a thing to see" (Smith 1996). Echoing the imaginative forays I had as an adolescent at the Spring Valley dump, others living in proximity to comparable piles of coal refuse also found the spaces of waste a source for play, conceding that, "On the prairie, even a good-size pile of rubble offers the opportunity not only for a lofty vantage point but also for adventure" (Smith 1996). Whether offering the opportunity to watch a car roll down the slope of pink shale or affording preteens the chance to race down the hillside to see who gets to the bottom first, the pile of industrial waste offered a space and performance of play, subverting the practices and consciousness that structure waste as something absent, dirty, and ideally removed from human contact.

But as some residents used the mounds of mining waste as spaces of recreation, others, specifically business and government officials, saw these dirty configurations of play as detrimental to the city's economic welfare. Arguments to remove the slag piles were aligned with beliefs that remediating the waste sites into traditional tourist and recreation areas could provide needed economic benefit. Subsequently, the pile of pink shale that once stood at the eastern border of Spring Valley has since been diminished to half of its size, with grass and shrubs now covering the remaining pink slope. As the 2021 Visitor's Guide authored by the Bureau County Tourism Committee describes, Coal Miners' Park, which the dump is now officially named, is "a sensory delight, offering spectacular views over the Illinois River with a variety of trees, fishing ponds, beauty and serenity close to shopping" (38). Through the practice of restoration, the green, parklike waste pile has become a space that fits familiar perceptions of the beautiful, where the purity and balance of the natural order has been reestablished thanks to human intervention.

Yet, as critics of practices of restoration make evident, the greening of contaminated spaces makes the dirty matter visibly absent and for many publics perceptually absent as well, even though the waste remains behind the grassy and green cover. Through such absences, publics perceive that the human master can clean up its mess, allowing the practices of industry and consumerism responsible for the dirty matter to continue. According to environmental advocates, revising these restoration practices so that publics encounter waste will thus lead to the acknowledgment of waste's presence and, moreover, invoke feelings of guilt and shame as publics bear witness to their complicity in such environmental destruction. Upon encountering dirtied and disturbed landscapes, publics acknowledge the damage that industrial and consumerist practices can cause, with the resulting affects, as many environmentalists envision, moving publics toward greater environmental care and environmental responsibility. And as purveyors of the environmental apocalypse advance, the publics' failure to cultivate these right affects and the appropriate actions will result in continued environmental destruction and to the eventual lifeless planet. While the intentions of the apocalyptic message should not be dismissed, the effectiveness of the approach and the capacity of the rhetoric to generate such environmentally conscious subjects can be questioned. Practices of restoration informed by modernity and that promise the return of the natural order remain a stable component of humanity's relation with waste, and industrial waste continues to spew into the air and water and toxins continue to circulate within the soil. Therefore, locating alternative modes of perception that may emerge from novel spaces of restoration and encounters with waste must become a part of contemporary environmental concerns. In other words, revising practices of remediation may be best advanced by the opportunities afforded by novel configurations and novel spaces of restoration. As I noted in the opening paragraphs of this chapter, for decades prior to its conventional restoration, the dump in Spring Valley provided meaningful encounters with waste. And while the various performances that emerged from this space may not fit easily into familiar actions that demonstrate common practices of environmental care, they, nonetheless, convey the inventive and creative resource afforded by encounters with such dirtied space. What becomes important for an inquiry into contemporary relations and encounters with waste is thus a consideration of the possibilities that emerge and how the perceptions and responsive actions to these encounters with dirty matter reveal the prospects of dwelling and, in the case of the dumps, playing with dirt.

My exploration is thus focused on the generative potential of these heterotopic spaces, of the potential for these configurations to invoke alterity in terms of human subjects, relations, practices, and ecological care. It is within these spaces of incongruity where novel practices of dwelling with waste may emerge and reveal alternative ways of being in a contaminated world. A significant amount of scholarship has explored heterotopia as specific places, as bordered and contained locations of multiplicity, incongruity, and otherness. Given the heterogeneous range of entanglements that characterize the contemporary, posthuman world, I extend the definition of heterotopia to also include geographically dispersed configurations, spaces that are constituted primarily by relations as opposed to fixed borders. Subsequently, I do not position or define space as an other to place; rather, I define space as the separation and distance between others, as a dynamic, perceptual, and geographic gap that may diminish or enlarge as heterogeneous bodies encounter dissonance and alterity. Heterotopia are thus material-discursive entanglements that may bring alterity and multiplicity closer, a perceptual proximity that calls upon human actors to reorient themselves to this changing arrangement of reality. In contemporary heterotopia, waste is an actor juxtaposed with many others, who collectively constitute material articulations of difference that disrupt the prevailing order. As such, heterotopia are rhetorical situations, prompting human actors emplaced within these configurations into some form of rhetorical performance that equips them with the means to respond and relate to the other. My definition of heterotopia shares affinity to the ecological move in rhetoric, a shift commonly assigned to Edbauer's (2005) reformulation of the rhetorical situation whereby "place becomes decoupled from the notion of *situs*, or fixed (series of) locations, and linked instead to the in-between enaction of events and encounters. Place becomes a space of contacts, which are always changing and never discrete" (10). Edbauer erodes the conceptual and constraining boundaries that have long been associated with space and the rhetorical situation. Subsequently, the human subject who is entangled within heterotopia that include the objects of waste is one who does not just *"do rhetoric"* as a response to the emplacement but rather finds oneself *"in a rhetoric"* [emphasis in original] (13). Within contemporary heterotopia, the human agent is thus engaged in both interpretative and productive rhetorical acts. The encounter with heterotopic alterity acts upon the individual, bringing to bear a particular material-discursive hermeneutic event that affords particular performances and opportunities to understand and act within one's entanglement with a dissonant, contaminated world. In this chapter, I want to further develop the definition of heterotopia and the relation between heterotopia and rhetoric and in doing so lay the foundation that will inform my analysis of the specific spaces of contamination I will discuss in the subsequent chapters.

HETEROTOPIA AS CREATIVE SPACE

Much of the scholarly effort to understand heterotopia starts with Foucault's 1967 essay *Of Other Spaces*, which contains Foucault's frequently cited definition of the term. Heterotopia "are something like counter-sites, a kind of effectively enacted utopia in which the real sites, all the other real sites that can be found within the culture, are simultaneously represented, contested, and inverted" (24). As scholars have noted, the explanation of heterotopia Foucault advances in the essay holds both possibilities and problems (Genocchio 1996; Johnson 2006; Topinka 2010). On one hand, the term offers a potential line of inquiry to locate and explore alternative sites and spaces of contest (Hetherington 1997; Saindon 2010; Wesselman 2013), and many scholars have taken up this possibility to investigate the way heterotopia as counter-sites offer a means for marginalized subjects to challenge hegemonic orders. Others, however, question the usefulness of this singular approach and voice concern about the merits of strictly associating heterotopia with alterity and resistance. "Among all the attempts to apply and make sense of the concept, there is a persistent association with spaces of resistance and transgression. Yet, curiously, this link is often asserted with little substantiation" (Johnson 2006, 81). These scholarly shortcomings, however, should not be solely traced to an ill-informed use of the concept but rather can be associated with Foucault's ambiguous and often contradictory definition and description of heterotopia in *Of Other Spaces*. Topinka (2010), for example, labels Foucault's definition as "unwieldly," noting that "Foucault does not offer a succinct or unproblematic definition of heterotopias" (57). The term's obscurity results in part from Foucault's wide ranging and sometimes contrasting examples, evident in his taxonomy that lists six principles of heterotopia. According to the examples Foucault cites, heterotopia include spaces of deviation, sites that change meaning over time, single spaces that exhibit heterogeneity, and spaces that can be publicly inaccessible. Commenting on the challenges presented by the wide range of examples and descriptions, Wesselman (2013) contends that "the diversity of Foucault's descriptions and examples thus simply makes it impossible to speak of heterotopia as a (single) "type" of space with a recognizable and stable set of features" (22). As a result of the definitional challenges found in the essay, some have suggested that the term's value rests not in attempting to identify sites or the characteristics that constitute heterotopia but rather in understanding the way certain configurations may operate as a site of difference.

> Only little can be gained by saying that a space *is* (or isn't) a heterotopia; the concept becomes more productive when looking at how a space (structurally and spatially) *works* [emphasis in original] as a heterotopia. Rather than use the term to simplify and reduce actual spaces, heterotopias should be taken as heterogeneous pluralities with elements that can be diverse and different in nature—and Foucault's concept serves to analyze which of those engender otherness, and how. (25)

Following Wesselman's line of reasoning, the potential of the concept rests not in further adding to what Johnson (2006) has labeled as the "dazzling variety of spaces" (81) that serve as illustrations of heterotopia but rather using the concept to shed light on the means by which the configuration of objects can engender otherness by representing, contesting, and inverting prevailing orders. Genocchio (1996) adds further clarity, suggesting that Foucault's work prompts us toward a line of thought in which

> we scrutinize and question the implications and possibilities of the slips, exceptions, oddities lurking at the very limits of the system that defines for us what is thinkable, sayable, knowable. The heterotopia is thus more of an idea about space than an actual place. It is an idea that insists that the ordering of spatial systems is subjective and arbitrary in that we know nothing of the initial totality that it must presuppose. (43)

Viewing heterotopia in terms of their potential, a capacity rooted in the space's heterogeneity, thus moves us to recognize the existence of alternative ways of understanding and acting within material-discursive realities, including different ways in which we spatially and discursively perceive and enact human relations with waste.

If heterotopia may potentially disturb the prevailing order or constitute an alternative order then specific attention should be directed at the means by which these configurations achieve such effects. Insight into how heterotopias may do such work can be enhanced by turning to other works of Foucault, most notably the opening of *The Order of Things*. Most pertinent to this elaboration is the mention by Foucault, of a passage in Borges,

> a passage that quotes a 'certain Chinese encyclopedia' in which it is written that 'animals are divided into: (a) belonging to the Emperor, (b) embalmed, (c) tame, (d) sucking pigs, (e) sirens, (f) fabulous, (g) stray dogs,

(h) included in the present classification, (i) frenzied, (j) innumerable, (k) drawn with a very fine camelhair brush, (l) *et cetera*, (m) having just broken the water pitcher, (n) that from a long way off look like flies.'" (Foucault [1971] 1994, xv)

In *The Order of Things*, Foucault remarks about his humorous response to the list, a reaction fostered not just because of the "oddity of the unusual juxtapositions" that constitute the list but also due to the "fact that the common ground on which such meetings are possible has itself been destroyed" (xvi). In this example provided by Foucault, heterotopia entail not just the spatial configuration of otherness but also the disordering of discursive practices made possible by the juxtaposition of the incongruent. As Foucault explains: "What is impossible is not the propinquity of the things listed, but the very site on which their propinquity would be possible" (xvi). Heterotopia thus encompass perceptual space by which human resources can be drawn upon to generate some congruity and relationality among seemingly discordant objects.

As more than just juxtapositions of material objects, heterotopias are also perceptual, instances when "the juxtaposition of heterogeneous elements is so incongruous and disruptive to our normal sense of order that we are unable to realize such perversity within a coherent and familiar domain" (Genocchio 1996, 37). The incongruity causes a moment of perceptual disruption, leading to efforts to reestablish the existing order by displacing those items that may be out of place, to emplace the incongruity within familiar habits of thought, or to generate new relations among the elements constituting the novel arrangement. Borges' incongruous list reminds us of the human propensity when responding to misplacement, even when that disorder takes the form of a discursive list. When encountering the list, we initially seek to revert to familiar interpretative frames, eliminating some items in the list to reestablish the familiar and proper categories and associations. The proclivity to draw from the familiar may restore mental equilibrium but neglects the creative potential made possible by incongruity and disorder. In other words, the emergence of novel configurations, including heterotopia, may not always lead to new and more ethical practices. "An Other place can be constituted and used by those who benefit from the existing relations of power within a society as in the case of the establishment of the workhouse or prison as a place of Otherness that becomes a site of social control through the practices associated with it and the meanings that develop around it" (Hetherington 1997, 52). As I described in chapter 1, configurations in which waste becomes present and heterotopic illustrate the potential to unsettle social relations, evident in the way pollution of the nation's air and waters generated initial reconsiderations of certain consumer and industrial practices and established new practices to manage industrial waste. Yet contemporary heterotopia constituted through the juxtaposition of waste and humans have not exhibited the same degree of disruption and creativity, as industrial waste continues to be generated and continues to intrude into human activities. Subsequently, the incongruity that marked these initial juxtapositions has become reconciled, in part through the entrenched practices and perceptions that characterize risk management. Moreover, fixed articulations advanced by environmental advocates that enforce the incompatibility of human encounters with waste and that use dirtied landscapes as invocations of shame fail to fully account for the perceptual reorientations made possible by these contemporary heterotopia. As a result, the appropriateness of these familiar habits of thought to the emerging spaces of contemporary contamination must be reconsidered while new habits of perception must be developed that enable the opportunity to understand and articulate the novel juxtapositions of humans and waste that have emerged in the contemporary epoch.

Subsequently, a study of heterotopia becomes entangled with spatial and ontological politics. Drawing from the work of Lefebvre, Ackerman (2003) explains that "Spatial practice is revealed through the study of spatial ordering and the instruments that maintain (or defeat) that order" (89). As spaces of otherness, heterotopia attempt to invert and defeat a prevailing order, yet such efforts may not always, as noted, materialize. Contemporary critical thought, particularly informed by poststructuralism, has centered its attention on power, as performed through human-driven discursive, economic, and political influence, as the primary instrument by which spatial practices and orders are maintained. New materialists, however, suggest that the nonhuman actor holds some influence, and the disruptive capacity that emerges through novel and incongruous juxtapositions speaks to such potential. Accounting for the dynamism of social change involves a recognition of the capacity of nonhuman objects to foster new configurations and relations that take part in a political process of (re)shaping our world. And as I noted earlier, one of the challenges to new materialism has been its hopeful promise that assigning agency to nonhuman objects will result in more ethical human subjects. Heterotopia, as co-constituted by nonhuman agents, challenges this assertion. The matter of inquiry with respect to the posthuman becomes then not the usurpation of agency, power, and ethics by the nonhuman but rather the capacity of the human to acknowledge, interpret, and, when possible, act upon the novel articulations advanced by nonhuman actions in ways that move beyond the familiar. Any movement toward novelty and an enhanced environmental ethic relies upon human perception of the capacities made possible by the entanglement in which the human agent is emplaced.

In advancing a role for matter as a recognized participant within a public, Bennett (2010) proclaims that "the political goal of a vibrant materialism is not the perfect equality of actants, but a polity with more communication between members," (104) a call which leads us to consider "how can humans learn to hear or enhance our receptivity for 'propositions' not expressed in words" (104). Ethical action becomes tied to a perceptual proficiency, a capacity that allows the human subject to cultivate appropriate responses. We may grant a publicness and political role to matter but how do humans then interpret and act upon the spaces of incongruity and dissonance that result from matter's influence? Bennett suggests that we need to "devise new procedures, technologies, and regimes of perception that enable us to consult nonhumans more closely, or to listen and respond more carefully to their outbreaks, objections, testimonies, and propositions. For these offerings are profoundly important to the health of the political ecologies to which we belong" (108). How do we listen and respond to those articulations of difference? How do we interpret and act within spaces of dissonances and move beyond existing habits that have failed to provide the utopia of a pure and ordered planet or the dystopia of a planet in ruin and beyond repair?

If we are to become better equipped to dwell with waste, we are then tasked with developing a spatial literacy, a hermeneutic means by which to better acquaint the human with configurations of incongruity. Such new interpretive procedures involve enhancing our understanding of the way spaces and configurations may offer up unexpected and alternative ways of being in a world of crises with each human and material other and identifying how vibrant matter may limit the efficacy of familiar and even desired perceptions and actions. We must also come to recognize how our emplacement within contemporary heterotopia may require a shift in our understanding of ethics and environmental care. To develop this hermeneutic proficiency, we need not start from scratch, however. Rhetoric has long been concerned with methods of influence and the tasks of listening and responding, of perceiving and acting. We can thus draw from rhetoric's intellectual foundation and its continuing attention to the way space shapes bodies and practices to better understand the influence of contemporary heterotopia and the possibilities for developing novel, relational perceptions and performances as a means to exist within dirtied spaces.

RHETORIC, SPACE, AND RESPONDING RELATING TO THE MISPLACED OTHER

While research in rhetorical studies has primarily been concerned with the effects produced by written and oral discourse, the field has directed some attention at other objects of influence, including configurations of material bodies. The volume of contemporary scholarship directed at sites of commemoration illustrates the vitality of the shift to other modes of influence, as scholars explore the way that the arrangement of various public spaces constitutes a collective set of beliefs, practices, and identities. Prelli (2006), for instance, discusses the arrangement of statues along Monument Avenue in Richmond, Virginia, noting the various shrines to Confederate leaders "together manifested in material form an internally consistent narrative, with each offering an exemplar of the Confederate leadership's valor and statesmenship" (14). The grouping of the statues generates specific effects, as the configuration "works to dispose the attitudes, feelings, and conduct of those who visit, dwell within, or otherwise encounter them." (13). The influence generated by such memory spaces is a product not just of the individual, material objects but can also be traced to arrangement, one of the classical canons of rhetoric. The arrangement of spoken and written words plays a fundamental role in shaping the effects of a text, making the means to properly configure one's discourse so as to achieve an appropriate force and response a central part of rhetorical pedagogy. A turn to the material and spatial moves the consideration of arrangement away from the organization of a written and spoken text to the potency of influence generated by specific configurations of objects and the organization of space. As Blair, Dickinson, and Ott (2010) describe, "places of memory are composed of and/or contain objects, such as art installations, memorabilia, and historical artifacts" (29), and these objects are placed into a specific order. "Memory places also prescribe particular paths of entry, transversal, and exit. Maps, arrows, walls, boundaries, openings, doors, modes of surveillance all encode power and possibility" (29). These paths move subjects among the objects in designed ways, trajectories that generate purposeful effects. The arrangement of objects is not restricted to spaces of memory, however. Prelli (2006), for example, explains that "gambling casinos are structured to stimulate pursuit of desires. Cathedrals are designed to inspire awe and reverence" (13). Shopping centers and individual retail stores also arrange objects in particular ways in order to produce the desired consumer effects. And natural parks, by way of directing hikers' movements along paths and trails, similarly attempt to construct certain effects and encounters with the material objects of the wilderness (Senda-Cook 2013). Whether the space becomes constituted as malls, memory spaces, national parks, churches, or a myriad of other recognized sites, the arrangement of objects and the movement among these objects generate recognizable and spatially appropriate perceptions and actions.

More recently, a number of rhetorical theorists informed by new materialism have turned to the arrangement of space to challenge claims that rhetoric emerges separate from matter's influence on the human agent. Frequently associated with the ecological turn in rhetoric, the shift in granting influence to the other-than-human in the production of rhetoric is grounded in the idea that rhetoric emerges through *"qualities of relations* [emphasis in original] between entities, not just humans, that enable different modes of rhetoric to emerge, flourish, and dissipate" (Stormer and McGreavey 2017, 3). This shift grants rhetorical potential to configurations; subsequently, new relations, including ensembles of seemingly incongruent matter, may generate new rhetorics and perceptions. "To spatialize is not only to order the boundaries of a rhetoric, it is also to create the conditions of possibility from which 'something new' might emerge: new meanings, subjectivities, bodies, practices" (Ewalt 2016, 146). And the rearrangement of space is particularly important to fostering new perceptions and rhetorics. "Different materialities set the field of potential and condition diverse rhetorics' emergence from the broader environment. If that environment changes, so too does rhetorical capacity" (Stormer and McGreavey 2017, 19). The creative potential of rearrangements has been taken up previously by rhetorical scholars, with articulation theory serving as one of the primary means to understand the emergence of new meanings and rhetorics from reassembled discursive objects. And while articulation has largely been used to identify the potential of new arrangements of discourse and ideological terms, the concept also has applications to the rearrangement of matter and the reconstitution of space.

Articulation, according to Laclau and Mouffe (2001) is "any practice establishing a relation among elements such that their identity is modified as a result of the articulatory practice" (105) and has served to inform rhetorical studies focused on social change and social movements. DeLuca (1999a), for instance, draws from Laclau and Mouffe to illustrate how articulation shifts the understanding of the toxicity that environmental justice residents face from an environmental problem determined through the science of risk assessment to an instance of racial and class discrimination. Linking elements within the discourse of economic and racial discrimination to those within the discourse of toxicity modifies the definition and assessment of toxins. As a result, risks from toxins are no longer cast in objective and qualitative terms but rather as specific examples of racial and class oppression. DeLuca (1999a) contends that articulation theory enables advocates to exploit the gaps of the discourse of industrialism to move understandings of toxic waste from "the price of progress and the normalized costs of economic growth" to "examples of class discrimination, institutional racism, and corporate colonialism" (42). Articulation not only offers a way to redefine understandings of toxicity and risk so as to reorient a public's perspective, it also offers the opportunity to revise a public's subjectivity and to grant a public rhetorical agency and authority.

DeLuca (1999b) references the transformation of Lois Gibbs from "housewife to environmentalist" (339) to illustrate the way subjectivity emerges by way of bringing various elements from existing discourses into new relations and arrangements.

> While it is obvious that the discourses of environmentalism and activism enable Gibbs to perform the role of environmental activist, Gibbs's immersion in the discourses of mother and property owner (housewife) are also crucial to her constitution as an environmental activist. It is her responsibility and right within these discourses to care for her children and her home that prompt her to enact the discourse of environmental activism. (339)

Lois Gibbs's transformation to environmental activist may be, as DeLuca attests, a product of the linkages among various discourses. But her new subject position is also a result of the new performances and practices that become constituted through the entanglement of the spaces of motherhood, property owner, and toxic waste. Practices to maintain the cleanliness of one's property and to ensure the safety of one's family find themselves entangled with toxicity. The possibilities of articulating new subjects and rhetorics become manifest not just through novel linkages of discourse but also the novelty that arises through attempts to reconcile incongruous juxtapositions of performances and practices.

Scholars exploring the rhetoric of materiality, particularly those exploring sites of commemoration, offer additional examples by which spatial reconfigurations, including those constituted by objects and practices that are seemingly out of place, can be a means to foster new perceptions. For example, Prelli (2006) notes how including the statue of African American tennis great and Richmond native Arthur Ashe disrupts the perceptions and order generated by the series of Confederate statues. The presence of the Ashe statue "functions as a material synecdoche of African American emergence into metropolitan, regional, and national civic life—a synecdoche jarringly incongruous with the embodied Confederate story celebrating civic leaders who defended a cause that barred that political possibility" (14). Blair (1999) documents a similar disturbance generated by the juxtaposition of the Civil Rights Memorial with the various monuments to the Confederacy located within Montgomery, Alabama. The Civil Rights Memorial "stands nearly alone as a reminder of the 'other,' more recent history of Southern and national racism. It is an interloper in the commemorative context, but it is also readable as a challenge to it" (43–44). Echoing the tenets of heterotopia, rhetorical scholars contend that space, as constituted by the arrangement and relation of objects, is dynamic and multiple, consistently open to revision by way of disruptions caused by the inclusion of seemingly out-of-place objects.

But if heterotopia or any novel arrangement of objects are to provide the potential for new practices to emerge, then fostering a human capacity to perceive and respond to such novel arrangements of incongruity anew becomes essential to the process. The initial sense of incongruity brought about by the misplacement of an object or by the seeming incompatibility of items in a heterotopic list is, as Gadamer (1997) suggests, a common human experience.

> The lack of immediate understandability of texts handed down to us historically, or their proneness to be misunderstood, is really only a special case of what is to be met in all human orientation to the world as the *atopon* (the strange), that which does not "fit" into the customary order of our expectation based on experience." (318)

But the perceptual and responsive tendency to such dissonance is to draw from the familiar to reconcile the discord or to remove the disturbing elements to restore some sense of the prevailing order. Drawing from the notion of *atopos*, Reeves (2013), however, outlines the creative and generative capacity fostered by monuments and public memorials that run counter to the conventions of commemoration. Referencing Blair and Michel's analysis of the Vietnam Veterans Memorial (VVM), a material artefact often cast as subverting the genre of the war memorial, Reeves describes how the out-of-placeness of the memorial recasts commemorative perceptions and performances.

> As Blair and Michel make clear, not only do memorials like the VVM reject the traditional tropes and representational logics of public memory; they also urge us to call into question the social and "commemorative" activities that take place at sites of commemoration. Just as generic works of public memory tend to elicit a detached, consumptive encounter, more subversive works have the potential to disturb the audience's complacent, everyday interpretive composures. (310–311)

Encountering the strange and out of place situates the interpretive subject within a precarious stance. The strangeness of an out-of-place text or material object jolts a body from a state of complacency fostered through convention and into a quest for a "response that is grasping, engaged, and ultimately creative in its forging of interpretive common ground" (311).

Encountering the out-of-place thus becomes a perceptive, hermeneutic act. "By allowing ourselves to be drawn into the world of a countermonument, we have not embarked on a typical interpretive endeavor; rather, we have adopted a task similar to that of the orator, who must find a commonplace appropriate to the challenges at hand" (311–312). As Reeves explains, the interpretive act sponsored by *atopos* shares affinity to the production of a rhetorical text in that both performances call upon a perceptual proficiency to make sense of and inform an appropriate response to the discord.

Faced with the dissonance brought about by heterotopic, incongruous juxtapositions, the human subject may embark on a performance so as to "reencounter the world" (Reeves 2013, 313). Such a reorientation may "compel one to forge a common ground with the object of one's displacement" (314) and establish a novel relation with the (re)placed object by dismissing familiar habits of thought. Encountering the strange and reconciling one's emplacement within such spaces of incompatibility thus requires not just an interpretive proficiency but also an acceptance of one's vulnerabilities. "Affectability or "response-ability" requires that things be inherently vulnerable to one another. Vulnerability is not a state of being at risk but of being entangled, which requires being at risk in varying passive-active relations" (Stormer and McGreavy 2017, 13). One's ability to respond to and reconcile the discomfort brought about by the unfamiliar entanglement with the other may entail drawing from familiar commonplaces, a process that clearly lessens the subject's vulnerability since the familiar response may return objects to their expected place and reestablish previous relations. But the vibrancy of materiality may constrain the capacity of the human agent to draw upon the familiar and reestablish the commonplace and may afford novel affects and responses to the other to emerge. As a result, vulnerability and the extent of one's response-ability become shaped by the influence of matter and the configurations in which one is emplaced as well as one's willingness and proficiency to engage with alterity and a novel hermeneutic.

Endres and Senda-Cook's (2011) exploration of the way space was reconfigured in attempt to generate a new relation between bicyclists and automobile drivers illustrates the potential for novelty and change when encountering and reconciling the strange. They discuss the Critical Mass bike ride, an instance in which "city streets are temporarily transformed from lanes for car traffic to paths for which lanes serve less use. Cyclists travel in a pack, making their collective impact greater than the cars they varyingly follow, bully (as motorists sometimes do to cyclists), and mock" (270). Space typically allocated for automobiles and kept separate from bicycles becomes reconfigured, and the familiar order becomes disrupted by misplacing a significant number of bicycles and their riders on the highway pavement. The encounter with the misplaced other prompts the automobile drivers to revise their relation to bicycles, responding to and reconciling the out-of-place objects by acknowledging that the "bikes are traffic" (270). While the potential to change a driver's perceptions are made possible by way of the reconfiguration, so too is the chance of drivers resisting the encroachment of vulnerability and subsequently maintaining their familiar identity and perspective, one that sees bikes on the road as a nuisance. Any change in a driver's perceptions prompted by the unfamiliar arrangement of traffic must account for the human subject's capacity to become perceptive to and affected by the other. The concern guiding any inquiry into the workings of heterotopia then becomes not just understanding how incongruity is generated and identifying the emergent possibilities of creativity and novelty sponsored by the dissonant juxtaposition. Attention must also be directed at the processes by which these creative and alternative possibilities materialize and gain cultural traction. This line of investigation becomes especially important given the challenges to revising the human vulnerability to waste. In the contemporary configurations of waste, vulnerability must extend beyond associations with the risk presented by the vibrant matter to also include the openness to and acceptance of alternative relations to the matter that may in turn help propel a more timely and appropriate ethic of ecological care in a dirtied world.

ACKNOWLEDGING WASTE'S ALTERNATIVE, HETEROTOPIC INFLUENCE

According to Massey ([2005] 2015), the possibilities of reconfigured space to promote change cannot be easily predicted and assigned to planned, human actions. She instead assigns, with some caution, chance a role in realizing space's possibilities.

> It is in the happenstance juxtaposition, in the unforeseen tearing apart, in the internal irruption, in the impossibility of closure in the finding of yourself next door to alterity, in precisely that position of being surprised (the surprise which de Certeau argues is eliminated by spatialization) that the chance of space is to be found. The surprise of space . . . . But this is not unique to the postmodern city or peculiar to heterotopic spaces: all spaces are, at least a little, accidental, and all have an element of heterotopia. (116)

While the movement of humans may result in a chance encounter with an unexpected other, similar meetings can result from the movement of nonhuman material. When emplaced within such movements and resulting configurations, the human agent must develop the capacity to recognize the creative opportunities that become present. But such possibilities can only become manifest through the human subject's proficiency to interpret and respond to the novelty the situation affords. Heterotopia, including those whose incongruity is fostered by the presence of waste, are rhetorical situations, and acting within the contemporary epoch of contamination requires a proficiency to interpret anew the contemporary dirty situations one becomes emplaced within and to identify the novel performances and relations such spaces afford. And while industrial waste may be viewed as a rhetorical situation in the traditional manner, as the existence of an exigence needing resolution, contemporary configurations of dirty matter force us to reconsider the emplacement of the human within contaminated configurations and the contemporary rhetorical situations humans are thus emplaced within.

Bitzer's (1968) article, "The Rhetorical Situation," has long been recognized as a starting point by which scholars have deliberated about the emergence of rhetoric, with numerous elaborations, including by Bitzer himself, aimed at furthering the initial work. Central to Bitzer's initial view is that "a particular discourse comes into existence because of some specific condition or situation which invites utterance," (4) a prompt he refers to as an exigence. Bitzer's definition of exigence connects us to the misplaced other and the influence of the out-of-place and the incongruent to invoke discourse. "Any exigence is an imperfection marked by urgency; it is a defect, an obstacle, something waiting to be done, a thing which is other than it should be" (6). The presence of waste has long been defined as an exigence, and the resulting legislative and advocacy efforts that occurred in the 1970s and the efforts to dispose of the dirty matter away from human encounters clearly illustrate the way waste's presence served to generate rhetorics. Yet the continued resistance of waste to be removed and the failure of the legislative and advocacy efforts to eliminate the dirty matter and spaces of contamination force us to reconsider waste's influence as an exigence. The incongruity of waste's presence remains, yet the familiar responses fail to adequately dispel the resulting dissonance.

According to Bitzer, various constraints act upon the rhetorical agent when inventing a response to an exigence. "Every rhetorical situation contains a set of constraints made up of persons, events, objects, and relations which are parts of the situation because they have the power to constrain decision and action needed to modify the exigence" (8). Rickert's (2013) new materialist rhetoric revises the rhetorical situation so that material objects and the relations among them must also be recognized as shaping rhetorical action. "We should begin to see the environment not simply as the location where information shows up or as the backdrop where human cognitive activity plays out but as an ensemble of material elements bearing up, making possible, and continually incorporated in the conducting of human activity" (93). In the case of those heterotopia that become manifest through movement of nonhuman actors, the possibility of alterity and novelty takes shape because of the newly formed human subject who is willing to encounter and respond in novel ways to the vulnerability invoked by heterotopic dissonance. The stable human rhetor is central to Bitzer; a situation may shape the rhetoric that is produced, but the environs have no influence in transforming the human subject. New materialist rhetoric, however, sees entanglements as productive not just of a rhetoric but of the subject as well. Subsequently, becoming entangled in an environment can itself be considered a rhetorical act since being in such relations influences not just the production of rhetoric and meaning but also the production of the bodies that are assembled and that establish some connection with each other.

> No "subject controls" what occurs; rather, actions emerge as willed by the situation precisely because there are no discrete subjects absent their relations and connections. However, the environs here are not just a material reality to which we adapt or a material situation that somehow "determines" us. Instead, the environs *enable*, but they enable *inclusively* [emphasis in original] of human beings insofar as human beings take shape within the environs. (Rickert 2013, p. 93)

Again, it is waste's vibrant persistence and the ineffectiveness of familiar rhetorical responses to waste matter that mark contemporary heterotopia of waste as different from previous iterations of rhetorical situations that included the dirty matter. A contemporary presence of waste affords the potential to not only generate new rhetorics but, just as important, to also generate new relations with dirty matter and resulting subjects equipped with novel perceptions that inform and enact a new ethic of ecological care.

Miller's anecdote pertaining to her attempts to address an intruding snake within her home provides a pertinent example to clarify the new materialist revision of the rhetorical situation and its applicability to understanding the contemporary heterotopic configurations of waste that are central to this study. The snake, clearly out of place within her home, became a part of the space in which Miller was also entangled, an arrangement that also included bookcases, houseplants brought in to escape the October cold, and a range of other objects. Guessing that the snake had ventured inside with the plants, Miller interpreted her surroundings so as to generate several actions to address the intruder and persuade the snake to leave. In generating these attempts to address the incongruity, Miller acknowledges that the assembled objects shaped her response:

> Whatever rhetorical agency I had was influenced—shaped, enabled, limited, constrained—by a highly particular convergence of things, practices, conditions, conventions, and beliefs in that place and time. The tin of water, the color of the carpet, the heaviness of the books and bookcases—all were part of the ambient configuration to which I, at least, responded, consciously or not. The snake can be cast not as the object of my appeals but as the initiator of the exchange. After all, I responded to the snake just as it did to me. What led me to respond to the snake as I did? How did I come to have available the particular means of persuasion I tried to employ? (Walsh et al., 2017, p. 458)

Miller's description of the way the matter configured within the space served as objects of influence to move the snake clearly echoes the new materialist version of the rhetorical situation described by Rickert. While her actions were constrained by the available material resources, her choices were not determined, as she admits to inventing several different efforts all designed to remove the intruder.

But while the configuration Miller encountered may have provided a variety of options to address the incongruity, these actions were, nevertheless, rooted in the singular intention and desire to remove the snake from the premises. In this regard, Miller's attempt to eliminate the snake can be compared to a public's familiar actions to remove and displace industrial waste, a similarly unwanted material intruder. Yet one interpretative path not recognized by Miller's anecdote consists not of eliminating the snake but of rather having to accept the snake's membership within the entanglement and the nonhuman's capacity to resist any and all responses to remove it. What if the snake exhibits an equal dose of cunningness, an ability to also draw from the available resources within this same entanglement and move on its own so as to defy all human attempts to direct its movement? What if the snake, like industrial waste, maintains a presence regardless of human actions? How might we perceive such an entanglement and how might the assemblage of the configuration and the dissonance caused by the unwanted object's continual presence reshape the subject's relation to this other and offer some means by which to dwell with this incongruity? The rhetorical situation is thus no longer marked by an exigence defined as the presence of the unwanted object and the attempts to remove it. The contemporary exigence is instead how the human subject may respond to the persistence of the other of waste and the resulting incongruity and dissonance such entrenchment provokes. The rhetorical situation thus becomes redefined not as an effort to remove waste and lessen human encounters with the material but rather as a means to enable the emplaced human subject to perceive and act upon and within the incongruity. It is in those different spaces, in the heterotopia that enable an encounter with the *atopos*, where we may be able to locate new perceptions and where new normative questions arise pertaining to the way we dwell in the contemporary contaminated world. It is in those exceptional rhetorical spaces that afford the opportunity to investigate the possibilities for new perceptions and new ecological, rhetorical performances.

Such a reorientation moves us toward thinking anew about the prospects of restoration, for example. The greening of the dump in Spring Valley that I described in the opening pages of this chapter can be seen as exemplifying the familiar and traditional practices related to restoration. And as I also alluded to, such practices are frequently dismissed by environmental advocates, who proclaim that such efforts mask the visibility of dirt and the practices that have generated the material. So what if, as with Miller's snake, the pile of pink shale and industrial refuse were left visible and not converted to the sanctioned site of play, beauty, and recreation? On one hand, the community performances of rolling automobiles down the lifeless slag pile clearly contradict practices rooted in environmental care and may even be seen as somewhat apocalyptic. This particular encounter with waste, at least when viewed through conventional habits of thought, clearly does not promote a better ethic of ecological care. In contrast, greening the space, even if we acknowledge its hidden ambition to hide the ecological damage, seems to offer performances and a corresponding ethic more aligned with those traditional values aligned with environmentalism. But watching cars slide down the waste slope was not the only performance enacted on the pile of waste. The hill also provided, even in its dirtied state, the same possibilities afforded in its current clean, green condition.

Many times, myself and others climbed the slag pile and, echoing the institutional language used to describe the current institutionalized Coal Miners' Park, gazed upon the spectacular views over the Illinois River and the beauty and serenity that surrounded the waterway below. Our gaze may have made available by what some see as destructive and anti-environmental behavior; yet it also provided the same opportunity to attune with the natural environment that is marketed today. This particular action did not prompt some transformative environmental moment within us, however; we did not reflect upon the industrial and consumer practices that led to this misplaced, pile of lifeless rock. Nonetheless, we experienced a significant and meaningful encounter, generated by the pile of slag and its relation to the other lively matter that extended below us and beyond the recognized borders of the dump. Those personal and culturally meaningful forays to the top of the dump to look out across the Midwestern horizon illustrate that perceptions of configured waste as useless space or as representations of our environmental guilt need not be the only means by which to accept waste's presence. In those moments, the dump, in its dirty manifestation, becomes a space that is neither idyllic nor apocalyptic but rather an accurate and appropriate configuration of the in-between space of the dirty and the clean and the resulting incongruity that constitutes contemporary bodies and spaces. Subsequently, if we can no longer escape dirt's presence and if those spaces of waste and restoration no longer serve to absolve our ecological shame, then we must inquire as to what possibilities are afforded by our entanglements with dirty matter and the effects engendered by these spaces and situations. In the chapters that follow, I explore a series of other contemporary heterotopia in which waste is present and that have generated cultural value and capital to locate the way such dissonant spaces may invoke novel perceptions and performances that reveal the prospects for dwelling in a contaminated world. Of particular interest is the response-ability of the human actor to generate and act upon new perceptions and practices afforded by these contemporary heterotopia and the way that such novelty reflects, contests, and inverts those familiar ethics and habits of thought that have informed the familiar relation between waste and environmentalism.

3

✢

Beyond Sustainability

*Relationality, Uncertainty,
and the Responsible
Posthuman Environmental Public*

Similar to some other industrial objects and structures, the smokestacks of the Tampa Electric Company's (TECO) Big Bend Power Plant have taken on an iconic regional status and have come to play a central role in defining the landscape on the eastern shore of Tampa Bay. Yet these smokestacks generate more than just visual representations of industrial prowess and environmental degradation. From November through March, the smokestacks serve as a visual compass point, directing tourists toward the power plant to witness the seasonal congregation of West Indian manatees as they gather within the discharge canal that borders the southern end of the facility. During these winter months, the water temperature within Tampa Bay and other nearby waterways can drop to levels that threaten the health and welfare of these mammals, who are dependent upon warm water to sustain their body temperature. As a result, the manatees seek refuge in the facility's discharge canal, where water used in the process of generating electricity is released back into the canal at temperatures warmer than the water in Tampa Bay. In many other spaces, such industrial effluent is defined as thermal pollution since the warmer water endangers the livelihood of biota populating the discharge area. But in the case of the Big Bend facility and other power plants in Florida, the effluent becomes redefined due to its role in ensuring the survival of the Florida manatee.

The particular configuration of industrial effluent and species survival at Big Bend exemplifies the incongruous juxtapositions that characterize contemporary heterotopia. Carr and Milstein (2021) identify Big Bend as a "contradictory space," and contend that paradoxical and heterotopic spaces such as Big Bend offer an important resource to investigate the way publics uncritically reconcile environmentally harmful practices and species preservation.

> We argue that it is within such explicitly paradoxical sites—wherein ongoing dominant practices clash directly with individual curiosity about, love for, and potential desire to protect the more-than-human world (Abram, 1997)—that research can shed revealing light on ways humans in Western/ized and or industrial/ized settings reconcile awareness of and care about anthropogenic environmental destruction with the daily and structural practices of extraction, overproduction, overconsumption, and waste that drive those environmental crises. (184–185)

But as Carr and Milstein describe, the dissonance inherent in the novel juxtaposition that constitutes Big Bend fails to perceptually or ethically disturb the visiting public. They assign publics a share of the blame for such interpretative malaise, explaining that "the processes we reconcile the incompatibility of dominant daily practices and looming environmental catastrophes—including climate disruption, mass extinction/extermination and food insecurity—require collective inattention and related inaction on an immense scale" (184). The "willful collective blindness" that is enacted by these visitors "serves a profound ontological need to protect dominant definitions of humanness premised upon a human/nature binary that positions humans as separate from and inherently superior to 'nature'" (185). To disrupt the collective blindness, Carr and Milstein propose "using travel reviews to both counter and educate about dominant discourses, pointing to the obvious in front of everyone's eyes and envisioning different ways of human being and doing" (191). Encountering the alternative, critical discourse will supposedly transform the human subject by way of offering visions and promises of a world rescued from its apocalyptic future. The alternative discourse that becomes available through these reviews conjures a public "ceasing to be complicit in reproducing the invisible sphere" and blissfully marveling at what a "world looks like without spaces of invisibility, without anthropogenic environmental, species, and climatic devastation, without networks of destructive overconsumption, overproduction, and waste" (191). While Carr and Milstein identify a crucial line of inquiry pertaining to the publics' perception of contemporary paradoxical spaces, they, nonetheless, revert to the familiar practices of critical discourse and ecological shame and situate Big Bend within the limited binary frame of the purified and the apocalyptic to promote public attention.

Casting aside familiar, untimely rhetorics, I use Big Bend as an opportunity to explore the value of identifying alternative perceptual practices that emerge from the particularities of the incongruous juxtaposition and that better equip humans to understand their emplacement within contemporary heterotopia. Publics may perceptively make invisible the practices that generate the warm-water effluent to reconcile the space's incongruity, or, as Carr and Milstein assert, the ecological dissonance may be resolved by materially eliminating those practices. Regardless, in both cases, order is restored by drawing upon familiar practices to eliminate and remove the dirty matter. Instead of promoting a habit of thought that erases the visibility and material presence of waste, something I have argued is increasingly difficult in contemporary spaces, I want to propose an alternative means of perception, one that acknowledges, rather than disavows, the increasing entanglements and relations with the other of waste and the corresponding increase in dissonance and discord that constitute the dirtied spaces we now inhabit. A public's reliance on existing practices and familiar frames to inadequately perceive novel and heterotopic configurations and to make sense of their dissonance speaks to the central concern of this book and is illustrative of what Braidotti (2013) refers to as "one of the most pointed paradoxes of our era," (58) namely the "tension between the urgency of finding new and alternate modes of political and ethical agency for our technologically mediated world and the inertia of established mental habits on the other" (58). While the critical discourse advanced by Carr and Milstein exhibits the inertia of familiar perceptive habits that have long fostered environmentalism's relation to waste, I want to turn my attention in this chapter to how the engagement with the array of alterity that marks contemporary heterotopia can be used to contest sustainability, an additional and what has become a highly prominent frame of contemporary thought. In fact, I would argue that sustainability has become the primary means to reconcile contemporary ecological dissonance. As an enactment of "our shared cultural narratives of sustainability" (Mentz 2012, 586), the configuration at Big Bend posits the potential of mediating difference through the coexistence of electricity generation and species survival and moves visitors to "a fantasy about stasis, an imaginary world in which we can trust that whatever happened yesterday will keep happening tomorrow" (586). Difference exists and relations with the other occur in current definitions of sustainability but only through the capacity of the emplaced actors to function independently and to retain their existing identity.

As I will illustrate, this guiding principle of sustainability and its enactment at Big Bend can be eroded by shifting to the posthuman view of relationality and vibrant matter and the role of the other-than-human in constituting contemporary space. In this regard, the configuration of Big Bend affords the potential, following Braidotti (2019), to practice posthuman hermeneutics:

> Posthuman thinking is a relational activity that occurs by composing points of contact with a myriad of elements within the complex multiplicity of each subject and across multiple other subjects situated in the world. Thinking takes the form of cartographic renderings of embedded and embodied relational encounters. These encounters can be with texts, institutions, or with other concrete social realities, or people. (123)

When perceiving space as constituted through multiplicity and dissonance, the focus of an inquiry turns to the way difference constitutes the relation among the various objects. As Miller importantly points out in her discussion of the connection between actor-network theory and rhetoric: "Rhetoric, like sociology, is interested in associations—in identifications, communities, adherences, agreements. A rhetorician, however, wants to know something about the nature and quality of those associations, in addition to their number" (Walsh et al. 2017, 456). More than recognizing that heterogeneous agents act upon and establish relations with one another, "rhetoric wants to explain it, to understand the influences on decision, adherence, and action, the motives for 'making a difference'" (Walsh et al. 2017, 457). If sustainability operates through disassociation, through acknowledged, separate and stable existences between others, Miller points us to the importance of attending to the construction of associations and the prospects for a mutual coming together that involves matters of contestation, change, and negotiation.

The approach I advance will be used to pursue concerns raised by Benson and Craig (2017), who contend that sustainability as a practice and discourse is no longer viable; change and unpredictability better represent our current era. Sustainability is grounded in the assumption that humans "have the capacity to maintain some type of stationarity and/or equilibrium in the relevant systems" (Benson and Craig, 2017, 4), an assumption that runs against current understandings of complex systems such as ecologies in which "humans are one component of vast networks of complex SES [social-ecological systems]—systems that sometimes react in ways that humans did not intend or, often, could not have even predicted" (5). Mentz (2012) advances a similar claim, suggesting that instead of rhetorics based on stasis and that construct a world we perceive as constituted through human intervention, "we must learn to love disruption, including the disruption of human lives by nonhuman forces" (587). In the following pages of this chapter, I want to situate the potential of disruption inherent in heterotopia to not only contest the prevailing tenets of sustainability but also to explore the way incongruities fostered by waste's presence call upon us to identify the possibilities and challenges of moving toward and associating with the others we are becoming increasingly entangled with.

SUSTAINABILITY AND THE MODERN QUEST FOR STABILITY AND ORDER

Central to shaping the American public's environmental imaginary has been a persistent ideology and rhetoric that positions nature and culture as two disparate bodies, irreconcilable spatially as well as conceptually. For instance, writing about the nineteenth-century photographs of Yosemite taken by Carleton Watkins, DeLuca and Demo (2000) suggest that

> in picturing a nature apart from culture, Watkins was obeying the dictates of the nature/culture dichotomy central in Western civilization, wherein a nature out there ontologically divided from culture serves as a source of resources, artistic inspiration, spiritual awe, emotional succor, and so on. Viewing nature as pristine wilderness apart from humanity becomes cultural convention and environmental policy, evident in pictures of other natural areas, from the Grand Canyon to Yellowstone, and inscribed in the Wilderness Act of 1964. (254)

Through symbolic representations such as photographs, paintings, and written discourse, the spaces of the uninhabited American wilderness become a "profoundly human construction" (Cronon 1996a, 25), with nature constituted as something other and able to invoke experiences unlike those spaces populated by human civilization. One prominent response spurred by encounters with the nonhuman was the sublime, an emotional reaction of fear and awe generated by an object that "is by definition something one is not accustomed to, something extraordinary" (Nye 1994, 23), with Niagara Falls and its "offering the experience of a powerful natural feature of superhuman scale that inspires awe and fear" (Spirn 1996, 95) cast as the representative example. Yet the distinctiveness of these natural spaces and objects extends beyond their extraordinariness. At the close of the nineteenth century, the natural landscape offered an alternative experience to the "confining structures of civilized life" (Cronon 1996b, 77) that became more prominent as a result of the growing wave of urbanization. Industrialism, urbanization's counterpart during the late nineteenth century, likewise came to be opposed to natural spaces, as the processes of capitalism and economic expansion increasingly placed a costly burden on the health and welfare of the inhabitants of the nation's growing cities. Places tainted by human practices came to be represented as examples of disorder, especially when cast in opposition to the natural order and harmony of spaces untouched by human intervention, with nature seen as "a refuge from man; a place of healing, a solace, a retreat" (Williams 1980, 80). "The ravages of industrial society ultimately gave rise to the fetishization of remote, inaccessible, and marginal landscapes," inspiring the initial movement to preserve the sanctity inherent to natural spaces (Davis & Zanotti 2014, 602). As spaces to escape from the dehumanizing tendencies of urban life or as areas untouched by the practices of industrialism, the seemingly unspoiled, natural areas scattered across the American landscape became decidedly defined and valued by way of their distinction to those places touched by human hands.

Toward the close of the twentieth century, however, the separation between nonhuman and human spaces began to erode. The dissolution of an untouched and separate nature emerged due to increasing anthropomorphic influences and intrusions but also through an accompanying new rhetoric, one grounded in the ideology of sustainability. The impetus to rethink the relation between humans and nature is presented in the opening pages of *Our Common Future*, considered as the founding document of sustainability.

> From space, we see a small and fragile ball dominated not by human activity and edifice but by a pattern of clouds, oceans, greenery, and soils. Humanity's inability to fit its activities into that pattern is changing planetary systems, fundamentally. Many such changes are accompanied by life-threatening hazards. This new reality, from which there is no escape, must be recognized—and managed. (WCED 1987, 18)

The emergence of sustainability and its corresponding rhetoric were heavily influenced by the iconic images of Earth from the Apollo space mission, which constructed a perception of the planet as a "single entity, united, limited, and delicately beautiful" (Heise 2008, p. 22). As opposed to reinforcing the separation between culture and nature, "the thing that distinguishes sustainability is looking at systemic interconnections, and the idea that the elements should support, or reinforce one another in a reciprocal relationship" (Vos 2007, p. 335). Central to such reciprocity, however, is the primacy granted to the human agent. "Humanity has the ability to make development sustainable to ensure that it meets the needs of the present without compromising the ability of future generations to meet their own needs" (WCED 1987, 24). Sustainability thus does not limit or restrain human capacity; instead, it is an opportunity, through human innovation, to continue existing practices and even to further economic growth. "We have the power to reconcile human affairs with natural laws and to thrive in the process. In this our cultural and spiritual heritages can reinforce our economic interests and survival imperatives" (18). A primary means of mediating industrial practices and environmental care can be found in the contemporary faith granted to humanity's scientific and technical prowess.

The reliance on technocentrism as a means to mediate the dissonance between humans and nonhumans affirms a particular environmental ethic, one that proposes that human practices can produce less environmental harm through the continued development of innovative technological products. Such a stance leads "proponents and adherents of sustainability to see technology as the ultimate solution to the sustainability problem" (Vos 2007, 337). Techno-science not only suggests that environmental impacts can be abated; it also argues that industrial and consumerist practices can remain unchanged and thus reifies human mastery. Alaimo (2012), for instance, turns her critique of sustainability toward the significant influence granted to the human capacity to develop technologies that enable contemporary notions of ecological reconciliation to materialize. Drawing from Heidegger, Rickert (2013) voices a similar concern and directs his criticism toward the way that the acceptance of technology's role and the persistence of human hubris impede any opening for alternative ways to conceive of sustainability (249–250). While critiques of the reliance on technology have become prominent, other ways to conceive of the world and especially the relation and even dissonance between humans and nature have interjected themselves into discussions of sustainability.

For example, scholars have lately taken to situating human actions within larger spheres of influence, including entanglements with heterogeneous nonhuman actors that shape our being in the world. A recognition of the vibrancy of matter includes an acknowledgment of the potential dissonance that emerges as unusual spaces and reconfigurations become constituted through the uncertain movement and incursion of nonhuman objects, placing constraints upon humans to respond to the novel incongruity in established ways. Matter's agency, specifically its ability to generate new heterotopia and to constrain human responses to the resulting dissonance, thus challenges the central ideas of balance, stability, and certainty that characterize sustainability.

> In pursuing sustainability in the Anthropocene, humans are asserting that we understand our complexly changing complex world well enough to be able to successfully pursue continuous economic development and social betterment for a human population that is steadily and significantly increasing without compromising the environmental amenities that emerge from healthy natural systems, even though these natural systems are changing at multiple scales in ways that we cannot fully predict. (Benson and Craig 2017, 35)

Subsequently, we need a conceptual shift so as to adopt "a better framework for thinking about our evolving relationship to nature, one that encompasses change and unpredictability as our new normal" (Benson and Craig 2017, 47). Benson and Craig's concern is with the inability of current definitions and practices of sustainability to provide a viable means to understand and address the change and uncertainty that are associated with climate change. They offer resilience as an antidote to this uncertainty and promote the capacity of humans to adapt to new environments as the best means to rethink the human position within a changed climate. One of the central principles in their approach is to eliminate "stress on the system" by reducing pollution and decontaminating those current dirtied spaces (175). Benson and Craig cite "technology forcing" as one means of achieving this end, a process that entails enhancing efficiency standards to spark technological innovation (176). Such an approach mirrors the existing conversations about the role of technology in addressing environmental change, and while technology has produced some benefit in this respect the force of such a process has not been fully realized. In addition, Benson and Craig claim that decontamination will provide for the extra clean space needed for species to adapt, given "climate change is likely to outstrip, or at the very least severely challenge, species, and ecosystems' intrinsic capacities to adapt, even if those capacities are not already diminished by anthropogenic stressors" (176). Conservation, remediation, technological innovation, and policy regulation are all vital parts of the approach advanced by Benson and Craig, and they devote much of their discussion to regulatory reform. Yet for such an approach to generate force, some engagement with the public, particularly in terms of locating ways in which publics may perceive environmental issues anew and within a more appropriate habit of thought will become important. In the pages of this chapter that follow, I want to suggest how a public paradoxical space such as Big Bend, when encountered through the posthuman, heterotopic lens I propose, can be a means by which to move publics toward new perceptions that can better equip them for the uncertainty and adaptation that Benson and Craig forecast and that constitute the prospects for dwelling with waste. Furthermore, the approach I advance also furnishes publics with the requisite understanding of not just the potential to think and act anew in such contemporary configurations but also alerts them to the constraints and tension placed upon the contemporary public's response-abilities.

A MYTHICAL STABLE AND MODERN SUSTAINABLE PLACE

The West Indian manatee is a marine mammal whose range of existence is dictated largely by warm freshwater temperatures, resulting in a habitat currently confined to the Caribbean and southeastern United States. The Florida manatee, a subspecies of the West Indian manatee, was classified as endangered in 1967, a listing prompted primarily by the high mortality rates of the species caused by fatal injuries from collisions with human watercraft. The diminished manatee population has resulted in various regulations that mandate certain speeds in waters where manatees congregate as well as prohibiting watercraft altogether from certain waters where manatees gather to feed and reproduce. Adding force to these conservation efforts has been the public popularity of these marine mammals. With its significant cultural currency, the manatee has become an important object within Florida's tourism industry, with various manatee products, ranging from T-shirts to stuffed animals and magnets, available for tourists to demonstrate their fondness for the animal. In addition, numerous locations have been set aside as manatee sanctuaries and refuges for tourists and locals to view the mammal, with various websites promoting the best places for tourists to observe manatees. These structured encounters, as some argue, hold the potential to advance a public's awareness of the threats to the animal and in turn foster a particular ethic of preservation (Boley and Green 2016; Milstein, 2008). While Florida locations such as Blue Spring State Park and Crystal River afford the opportunity for a public to come to know the manatee in what are considered natural warm-water refuges (bodies of water fed by warm springs), a similar environmental ethic may be fostered in refuges that are fed by warm, industrial effluent. One of the more popular industrial warm-water refuges that offers public encounters with manatees is the Manatee Viewing Center (MVC), located adjacent to Tampa Electric Company's Big Bend Power Plant.

In this particular configuration, the objects and practices used to generate electricity and the objects and practices that maintain the preservation of an endangered species coexist, with water serving a key role in establishing the relation between the two, often opposing, practices. Water is withdrawn from the canal that runs adjacent to the power plant and is transported into the facility as a main element in the process of generating electricity. The water first enters a boiler, where the heat generated from the burning of coal raises the temperature of the water. Once the temperature increases to transform the water to steam, the pressurized water vapor is sent to the turbines where the force of the steam turns the turbine blades and produces electricity. The steam is then transported to another location within the facility where it is cooled and returned to water prior to its discharge back into the canal. As the water reenters the canal from the power plant, it is warmer, especially during the winter months, than the ambient water in the canal and in the bay, a difference that provides the manatees the needed warm water during the winter months.

While this novel juxtaposition helps ensure the health and welfare of the manatee, its incongruity is reconciled by familiar rhetorical frames. "Inside the MVC's environmental education building, colorful displays immerse you in the world of the manatee and its habitat. Others show how Big Bend Power Station generates electricity for the community in an environmentally responsible way" (TECO 2021, para. 3). Through common material-discursive rhetoric, the MVC promotes a familiar understanding of sustainability, particularly the ability of industrial practices to proceed undisturbed given the perception that these practices do no harm to the natural environment, and in this case actually benefit the nonhuman. Yet this familiar habit of thought subsequently neglects the contemporary ecology and geographically distributed agents that constitute Big Bend and their entanglement within other spaces and practices within Florida. The reliance on the familiar leads to the perceptual myopia of the modern tourist subject, a hermeneutic lapse that leads to an inadequate understanding of the prospects for dwelling within contemporary heterotopic and paradoxical space. The familiar rhetoric that constitutes Big Bend can easily be criticized as one additional example of the long list of corporate greenwashing practices; likewise, as Carr and Milstein (2021) propose, the MVC can also be criticized as a failure to invoke public attention toward the environmental damage of the continued reliance on fossil fuel use. These challenges, long a part of modern environmental advocacy, nonetheless offer little to equip publics with the requisite apparatus to better accommodate the position of humans within heterotopic entanglements of waste. Eliminating carbon from the space, while aiding the climate crisis, does not eliminate the heterotopic quality of the site; instead, it further reveals the challenges of human action within posthuman entanglements and the need to more fully attend to the prospects of contemporary ecological care. Before further developing this line of thought, however, I want to provide a brief sketch of the way that the material-discursive is used to constitute a familiar understanding of space and the way the configuration of the MVC generates an enactment of sustainability.

To move the public toward reconciling the paradox of Big Bend, the material-discursive encounter proffered by the MVC invokes the similar experience of interacting with the manatees as one would find in the spring-fed warm-water refuges. Such a mirrored reality is generated by emplacing objects within the MVC similar to those found in places where the public has access to view manatees in spring-fed warm-water refuges. The configuration of the MVC can thus be characterized as an effort to arrange space largely by way of immutable mobiles, specifically objects that move into new arrangements and bring their past histories and actions into this novel configuration.

> An immutable mobile, says Latour, is something that moves around but also holds its shape. Indeed, in this way of thinking, it holds it shape in two importantly different ways. On the one hand, it does so in physical or geographical space. On the other, it holds its shape in some relational and

possibly functional manner where it may, to say it quickly, be imagined as a more or less stable network of associations. (Law and Singleton 2005, 335)

As immutable mobiles, objects can move from one entanglement to another yet retain the same shape, agency, and potential, even as they become parts of new entanglements. In other words, the (re)placement of an object within a new configuration does not always alter that object's meaning or potential. For example, one of the central objects in the MVC is a boardwalk that runs adjacent to the discharge canal. The boardwalk allows visitors the opportunity to walk along the discharge canal and observe the manatees in the water. The boardwalk, by way of the spatial position next to the water and to the manatees, translates the space and visitor experience into a wildlife, educational encounter because of the boardwalk's associations to other places where wildlife can be similarly encountered. For example, boardwalks and the assemblage in which they are entangled in Blue Spring State Park perform the similar translation and offer the similar tourist encounter with manatees.

> Blue Spring State Park is unique because visitors can view manatees in the crystal-clear spring water from the boardwalk, which stretches $1/3$ of a mile from the St. Johns River to the headspring. The boardwalk provides a wonderful opportunity for safely observing the manatees. Visitors can see manatees socializing and mothers nursing their small calves, without influencing the manatees' natural behavior or bothering them. (FDEP 2021a, para. 4)

A second boardwalk at the MVC provides public access through the tidal area adjacent to the discharge canal, offering tourists the opportunity to move through another particular Florida landscape populated with local plant and animal species. Here as well, an environmental ethic that mirrors those prompted by a similar configuration within natural parks is generated by the placement of the boardwalk and hiking paths. Through these familiar objects and their arrangements, tourists can draw on familiar habits of perception, associating the ecology of the MVC with the spaces of sanctioned wildlife encounters emplaced in commonly referenced natural environs.

In addition to the boardwalk, genre serves as another object within the MVC that also exhibits the characteristics of an immutable mobile and further helps configure the space as a performance of sustainability. In the MVC configuration, genre takes up the purpose of conveying educational information, focusing largely on providing content about the biology of the manatee and the need to preserve the species. Moscardo, Woods, and Saltzer (2004) list a variety of means by which visitors to wildlife refuges are educated about the animals they view, including "interpretive signs, models, brochures, guides, demonstrations and shows, video, audio commentary, computers, and books" (236). The genres emplaced at the MVC not only correspond to those common to wildlife refuges but also mirror those genres found in the spring-fed warm-water refuges where the public may encounter manatees. "An entity such as a text or a device is immutable when its elements do not change and the relationship between them is not altered. It holds itself stable wherever it goes. And it is mobile because, from the point of view of a regional topology, it displaces itself from one place to another" (Mol and Law, 650). So while the boardwalk moves from the natural manatee refuges to the MVC and performs the same function in constituting the space in familiar ways, genre, as a means to convey educational content, also moves in a similar fashion and becomes (re)placed to invoke common relations among the disparate practices of electricity generation and manatee survival.

While the subsequent emplacement of various objects such as boardwalks and genres configures a space that promotes performances of ecological tourism, other objects emplaced within the MVC clearly challenge these particular enactments and perceptions. Unlike the spring-fed warm-water refuges, the configuration of the MVC includes smokestacks and other objects involved in the production of electricity that are readily present as one walks on the boardwalk viewing the manatees. At certain times, white emissions emerge from the smokestacks, constituting a configuration that runs counter to that found in natural areas where industrial emissions are commonly absent. To reconcile the incongruity, the MVC draws upon the familiar rhetoric of sustainability, discourse readily apparent in various interpretative panels located within the MVC. For example, one panel draws from the common notion that technology affords the opportunity to generate the balance between human practices and ecological systems. "Scrubbers reduce sulfur dioxide emissions and are part of the process that produce the white plumes you see coming out of the Big Bend Power Station's stacks. The white plumes form when the hotter wet stack gases, cleaned by the scrubbers, come into contact with the cooler air in the environment" (TECO, 2016). Through familiar discourse, TECO constructs itself as the modern corporate citizen who, through technological resources and innovation, can achieve ecological, commercial, and industrial balance.

The configuration of the objects, including the boardwalk, hiking paths, scrubbing technology, warm water, genre, and discourse within the geometric space of the MVC all attempt to influence a visiting public to adopt an understanding of the congruity between electricity production, industrial effluent, and manatee preservation. Yet perceiving the space through the frame of sustainability neglects the range of vibrant actors that constitute the heterotopia, leading publics to inadequately perceive the ecology of such contemporary spaces and juxtapositions of waste.

Studies have concluded that the relation between industrial effluent and manatees is far from stationary (Laist and Reynolds 2005; Runge et al. 2017; Flamm, Reynolds III, and Harmak 2013). Even the U.S. Department of Interior acknowledges the mutability of the industrial warm-water refuge, noting in its documentation to support the reclassification of the manatee from endangered to threatened that "a majority of the significant power plants used by wintering manatees have been repowered and have projected lifespans of about 40 years" (USFWS 2017, 16674). Understanding the ecology of spaces such as the MVC entails a recognition that the space is constituted by more than relations among warm-water effluent, smokestacks, boardwalks, genre, and those other objects visibly and materially contained within the geographic borders of the MVC. Numerous other actors whose presence is not immediately visible to tourists are also influential in shaping the configuration.

> Studies suggest that when individual power plants might be retired is unclear. Such decisions involve proprietary economic forecasts made by individual utilities based on dynamic hard-to-predict factors, such as the future cost and availability of alternative fuels, competition with rival power companies, projected trends in electricity demand, and the cost of renovation versus building new plants. (Laist and Reynolds III 2005, 281)

In fact, the actors present within the configuration of the industrial warm-water refuges are similar to those Bennett (2010) claimed had influence in the 2003 electricity blackout that affected parts of North America: "To a vital materialist, the electrical grid is better understood as a volatile mix of coal, sweat, electromagnetic field, computer programs, electron streams, profit motives, heat, lifestyles, nuclear fuel, plastic, fantasies of mastery, static, legislation, water, economic theory, wire, and wood—just to name a few" (25). Reconciling the incongruity of heterotopia through the discourse of sustainability is deceitful and inappropriate to contemporary publics not just because of the way the approach reifies detrimental corporate behaviors and identities; a public's complacency and ignorance rest more importantly in its inability to recognize the fallacy of balance and stability as it applies to the dispersed relationality of contemporary space and the lack of a readily available perceptual apparatus to better interpret the contestations and influence among a range of actors whose entanglement constitutes these contemporary configurations. Profit motives, government regulations, and, perhaps most influential, the improbability of an endless availability of a carbon-based fuel central to the processes whereby warm water is discharged into the canal all stand as highly influential actors who offer a persistent challenge to the integrity of the space. In this regard, the USFWS offers an alternative rhetoric to that encountered by the visitors to the MVC, contending that "addressing the pending loss of warm water habitat from power plant discharges remains a priority activity needed to achieve recovery" (USFWS 2017, 16679). Yet this discourse is not readily available at the MVC; rather, the discourse and alternative understanding of the heterotopic configuration are accessible through lengthy and formal government reports that require a degree of proficiency to access electronically. The government discourse affords the opportunity to move publics away from sustainability's guiding notion of stability and also offers the opportunity to address the myopic perceptions that limit the modern, human subject from understanding and attending to the complexity of contemporary interactions with the presence of waste. Consideration, however, must be given to how to make such discourse more readily available.

While the various actors entangled within the configuration of Big Bend may eventually exert influence to remove the industrial effluent, any subsequent absence of the waste and the detrimental practices by which it is produced, nevertheless, will lead to the enhanced presence of waste in other spaces in central Florida. For example, drawing on models predicting a "post-power-plant-discharge" future for the manatees, the state and federal agencies overseeing the management of the manatee argue that the viability of the species rests in measures taken to protect "natural and nonhuman dependent sources of winter warm water" (USFWS 2017, 16680). Such steps would entail efforts to restore the viability of spring-fed warm-water refuges, waters that are currently threatened by pollutants, primarily nitrates, and by the diversion of groundwater to satisfy Florida's growing residential and tourist populations. Therefore, decontaminating Big Bend by eliminating the carbon-fueled processes and resulting effluent results in the need to mitigate the contamination that exists elsewhere. Doing so, however, involves accounting for the uncertainties and range of vibrant actors that have reconstituted these spaces.

In the case of Big Bend, research has yet to conclude that manatees who have migrated to the power-plant discharges for warm water would successfully move to the spring-fed refuges once the industrial effluent is no longer available. For example, Flamm, Reynolds III, and Harmak (2013) draw from existing research to note that humans currently possess limited knowledge of the manatees' movements and behaviors in relation to warm-water refuges. When warm-water sites have become nonexistent, manatees may remain at the site and die, move to other nearby locations that may offer some warm water, or migrate southward to seek warmer water (161). Subsequently, decontamination, as in the case of Big Bend, entails more than removing the waste. Encountering the eventual absence of the waste also entails understanding the ecological reorientations that have occurred due to waste's presence and the constraints human face to not only rid a space of dirty matter but to also, as Benson and Craig suggest, open up spaces for species adaptation. Benson and Craig maintain that preserving current ecosystems is central to adapting to climate change, but they also contend that it is crucial to promote the capacity of all the parts and actors entangled within ecosystems to "assemble and reassemble as functional ecosystems" (172). But these systems, due largely to waste's presence, have already become reassembled. As a result, eliminating the waste at Big Bends leads to definitional questions with respect to what preserving an ecology entails. Further, humans must also play a role in protecting and creating "as many corridors as possible to connect those habitats and ecosystems so that species can move to new ranges as they need to" (172). In the case of Big Bend, perception and action also turn to the human response-ability to assemble and reassemble the functional ecosystems that are the warm-water springs. As I will describe in the ensuing pages, achieving these ends and building a perceptual apparatus by which publics may better understand their response-ability within the vibrancy and uncertainty inherent in posthuman spaces become a challenging task for any future mode of ecological care.

FLORIDA'S TAINTED SPRINGS

While Florida is often referred to as the Sunshine State, it would be hard to discount the role of water in shaping the state's identity. The influence of water can be traced to its significant presence, not just in the miles of coastline that encase the peninsula but also in the unique number of first-order springs (large springs with discharges historically greater than 100 cubic feet per second) that can also be found in the state. "Thirty-nine of Florida's sixty-seven counties have springs, or include areas of land that contribute water to springs (known as springsheds)" (Harrington, Maddox, and Hicks 2010, 2). The springs are fed by the Floridan Aquifer, water that rests under all of the state of Florida as well as parts of Georgia, Alabama, and South Carolina. Water in the aquifer is stored within the various limestone caverns and crevices resting under the sand and thin soil that cover the surface of the state, while vertical fissures within the limestone afford the water the capacity to move upward to the surface. Velousia–Blue Spring, located in the central part of Florida and approximately thirty miles northeast of Orlando, is one of the state's first-order springs, and its meaningfulness can be found by way of its entanglement with various practices. Located within Blue Spring State Park, the spring helps enact several recreational activities, offering tourists and residents the opportunity to swim and kayak in the spring-fed waters. The spring also affords a much-needed warm-water refuge for manatees during the winter months, and in turn, the spring serves as a place in which tourists and locals can encounter the manatee in what many would consider the species' natural habitat. As such, it provides an alternative space to the industrial warm-water refuges and is considered a crucial ecology to preserve given the predicted demise of sites such as Big Bend. Similar to the experience at the MVC, visitors to Volusia Blue Spring can view manatees from a boardwalk that runs along the spring and is bordered by the forested growth of trees on each side. It is a place that exhibits familiar traits of the natural, visually removed from the urban, human practices that rest outside of the park's borders.

While the spring is emplaced within the noted recreational and biological assemblages found within the borders of Blue Spring State Park, it has also become entangled with other practices and spaces in the state that have led to Volusia Blue Spring, along with twenty-three other springs, as being labeled as "impaired for the nitrate form of nitrogen" (FDEP 2018, 9). Volusia Blue Spring is one of thirty Outstanding Florida Springs (OFS), given that designation through the 2016 Florida Springs and Aquifer Act, legislation designed to protect and restore the springs and their springsheds (FDEP 2018, 14). As the Florida DEP 2018 report acknowledges, Volusia Blue Spring and thirty other OFS "are impaired by nitrate, which in excess has been demonstrated to adversely affect flora or fauna through the excessive growth of algae. Excessive algal growth results in ecological imbalances in the spring and run and can produce human health problems, foul beaches, inhibit navigation, and reduce the aesthetic value of the resources" (14). To lessen the amount of nitrogen and nitrate within the springs, the DEP has established Basin Management Action Plans (BMAP), "a water quality restoration plan that identifies strategies and projects to reduce sources of pollution to a waterway" (Donaldson 2019, para. 1). A critical task of the BMAP involves identifying the sources of the nitrates leading to the degradation of the spring, with such influences spatially identified within what is referenced as a primary focus area (PFA). For Volusia Blue Spring, the PFA or connectivity between the sources of nitrates and the spring "comprises 108 square miles and encompasses portions of the City of DeBary, City of DeLand, City of Deltona, and City of Lake Helen; all of the City of Orange City; and a portion of unincorporated Volusia County" (FDEP 2018, 15). If publics are to better understand the space of Big Bend through an acknowledgment of the eventual absence of warm-water effluent from the configuration, then a similar spatial perception must be adopted to understand the ecological instability and the extent of distributed actors that affect Volusia Blue Spring and efforts to restore the manatees' access to this needed source of warm water. The improbability of eliminating the nitrates and removing all of the dirty matter from these springs likewise demonstrates the need for some new way of perceiving and being within this dirtied space so as to better understand the constraints upon and novel potential made available to humans in facing the challenges of restoration in a dirtied world.

SPACE AND SEPTIC SYSTEMS

Roughly ten miles southwest of Blue Spring State Park is the town of Deltona, a city that emerged as part of an initial development by the Mackle Brothers, who were responsible for a significant amount of Florida development during the 1950s and 1960s. In 1962, the brothers bought "17,203 acres and filed a planned unit development for a community of 35,143 lots" in southwestern Volusia County, a development that would be called Deltona (City of Deltona n.d., para. 5). This initial development plan set the foundation that guided the exponential growth of Deltona, a city which has grown from a population of 4868 in 1970 to an estimated 92,757 in 2019 (U.S. Census Bureau 2019, Deltona City, FL). Such a rapid rise in population in such a short period of time is not unique to Deltona, as Volusia County has seen similar exponential expansion. (U.S. Census Bureau 2019, Volusia County, FL).

At the time of Deltona's initial development, little urban infrastructure existed, with one of the more notable omissions being the lack of a public sewage system. As a result, human waste was remediated by way of residential septic systems (OSTDS), a common practice in residential developments in many parts of Florida during the mid twentieth century. A 2012 study contends that "approximately 2.5 million systems on record are currently in use by approximately thirty-nine percent of Florida's population" (Badruzzaman et al., 84). As the population of Deltona has increased and the development of residential areas has grown, so too has the number of septic systems located within the identified PFA. "Overall, there are currently over 26,000 OSTDS in the PFA on lots less than one acre, based on FDOH estimates" (FDEP 2018, 28).

Studies specific to identifying the source of excess nitrates in Volusia Blue Spring have concluded that septic systems are the primary cause of the increase.

> Based on the Volusia Blue Spring NSILT estimates and GIS coverages, OSTDS contribute approximately 39 % of the pollutant loading in the PFA. Irrespective of the percent contribution from OSTDS, DEP has determined that an OSTDS remediation plan is necessary to achieve the TMDLs and to limit the increase in nitrogen loads from future growth. (55)

Septic systems are known to effectively treat human waste, but the efficacy of these systems is tied to numerous factors. More common to rural areas with sparse populations and where the residences are spatially dispersed, septic systems can remediate human waste so that it presents little environmental impact or threat to human health. Yet in more populated areas and in spaces with a high density of residences, the practices of treating waste by way of septic systems can disturb natural ecosystems because of the limited space allotted for the waste to be properly filtered. The difference in the efficacy of the process can be clarified by briefly explaining the process of treating human waste by OSTDS. In a septic system, the solid human waste is initially treated with fluids in a buried and enclosed tank, where the solid waste decomposes and settles to the bottom. The liquid effluent from this process exits the tank and moves to a drain field, a location within the residential property, where the effluent is dispersed into the ground with the intent of being treated through natural, biological processes. Nitrogen emerges from the decomposition process and is subsequently released into the soil and dispersed throughout the ecosystem. Within a condensed area consisting of numerous drain fields, the available biological processes cannot sufficiently address the amount of waste that subsequently disperses into the ground.

To address the migration of nitrates from residential septic systems, the state of Florida has enacted the Septic Upgrade Incentive Program. As explained by the state, the program "offers subsidies, only in designated areas within a county—identified and delineated by the Department as Priority Focus Areas (PFAs), in amounts up to $10,000 per system" (FDEP n.d., "Septic Upgrade Program," para. 2). Monies are to be distributed to residents to help defray the costs of installing new septic systems that are equipped to degrade the nitrates prior to the release of the effluent into the drain field. Nine counties and their respective PFAs which encompass the center of the state are included in the program, with watersheds that not only include various springs but that also include both coasts and those extending south into Lake Okeechobee. The program is thus not only an effort to address the presence of nitrates in various inland springs and that threaten the manatees' warm-water refuges; it is also a larger effort to address the growing presence of blue-green algae, which has been tied to the extensive number of OSTDS found in numerous other waterways throughout the state. In fact, the presence of the algae has become so significant that it has led to the formation of the Blue-Green Algae Task Force, a state agency whose goal is to "expedite improvements and restoration of Florida's water bodies that have been adversely affected by blue-green algae blooms" (Donaldson 2019, 1). The task force has identified that increased delivery of nutrients to Florida's water bodies is widely recognized as the primary driver of algal proliferation and subsequent degradation of aquatic ecosystems. Major sources of nutrients include, but are not limited to, agricultural operations, wastewater treatment plants, onsite sewage disposal systems and urban storm water runoff. (1)

As noted, new technology is being made available that will enable residential disposal tanks to more effectively treat the nitrates prior to their dismissal to the drain field, and funds have been made available to address the costs of replacing the older, inadequate residential systems. Yet, as of this writing, the funding provided by the state has been exhausted, leaving residents without any financial assistance and putting a halt on efforts to eliminate the presence of the nitrates that make their way into the springs through this specific practice (FDEP 2021b). In addition, regulatory requirements are being enacted that will move individual residents toward a renewed responsibility and awareness of the disposal of their bodily waste. While one component of the septic incentive program is to prompt the adoption of more innovative technology, the program also calls upon the need for continued maintenance and inspections of each individual septic system to ensure its regulatory compliance and effective operation. As suggested by the Blue-Green Algae Task Force:

> The task force recommends the development and implementation of a septic system inspection and monitoring program with the goal of identifying improperly functioning and/or failing systems so that corrective action can be taken to reduce nutrient pollution, negative environmental impacts and preserve human health. At present there is no requirement that conventional septic systems be inspected post-installation. (Donaldson 2019, 1)

The yearly inspections and any subsequent modifications that will be mandated are not, however, covered by the state subsidies. Subsequently, residents will have to pay the cost of the yearly inspections and any subsequent maintenance to the equipment. Because of the enacted program, individual citizens may no longer readily dismiss their association with and responsibility to waste. But the incentive program and the annual inspections also reveal the challenges to human efforts to dwell with waste, primarily in terms of the availability of a sufficient means of persuasion that can engage publics in attending to the responsibilities of living with dirt and the procurement of the revenue needed to develop, adopt, and maintain the technology and associated programs that will be a part of the act of dwelling. In the case of the nitrate pollution of the Florida springs and the association of OSTDS to the dirty matter, questions clearly emerge as to the means to move publics toward requisite actions and responsibilities.

For example, in public meetings discussing the incentive program with Florida state officials, residents of Deltona voiced skepticism as to how the waste from their current septic systems was affecting waters such as inland lakes and Lake Okeechobee (FDEP 2019). Studies conclude that excess nitrates exist within the waters of Volusia–Blue Spring and that the incursion of these nitrates results from the movement of waste material from residential septic systems; yet public comments at the forum expressed the lack of recognizable algae growth in waters surrounding Deltona. The presence of blue-green algae in other waterways and areas of the state and, more important, the circulation of dystopic images and discursive accounts of the algae in these waters that have populated the Florida press have carried significant influence in shaping an understanding of the state's water crisis. It is this rhetoric and its constitution of a particular dirtied Florida that have entangled the residents of Deltona. For example, images of the dirtied St. Lucie and Caloosahatchee Rivers, the two major waterways flowing from Lake Okeechobee and to the east and west coast of Florida, have come to dominate news coverage and subsequent public and political attention, especially after the significant disturbance caused by a significant growth of blue-green algae in 2016. The opening paragraph from a 2016 article by CNN that documents the fish kills related to the presence of blue-green algae not only corresponds to the familiar apocalyptic frame but also places the dirty matter in a space geographically and perceptually distant from the warm-water springs affected by the residents of Deltona.

> Florida may be the fishing capital of the world, but you'd never know it from the latest scenes around the state's Indian River Lagoon. Usually idyllic beaches, waterways and estuaries near the massive, biodiverse ecosystem along central Florida's Atlantic coast are littered with scores of dead, rotting fish; an estimated hundreds of thousands of them are floating belly up in brackish, polluted water as far as the eye can see. (Grey 2016, para. 1)

An article in the *The Broward Palm Beach New Times*, "Five Powerful Photos Show Massive Fish Kill in Florida" also describes the intrusion of nitrates and algae in similar rhetoric and focuses attention on the dirtied coastal waters of the state (Swanson 2016). The prominence of these images and the accompanying spatial orientation to Florida's dirtied waters can be traced to the ease by which fish kills and algae-laden water fall into familiar visual and discursive apocalyptic frames and the economic and cultural prominence granted to Florida's coasts. No similar dystopian rhetoric has been circulated that conveys the dirtiness of the waters around Deltona; and while some rhetoric has been generated to promote the restoration of the springs in relation to their importance to the warm-water habitats of manatees, this rhetoric has failed to garner the same level of public attention as the rhetoric associated with blue-green algae. As a result, the entanglement of the Deltona residents within regulations aimed at remediating Florida springs results more from the circulation and potency of the familiar apocalyptic rhetoric associated coastal infestations of blue-green algae and the cultural and economic influence of Florida's coasts. The relationality constituting the contemporary dirtied state of Florida and efforts to dwell with the matter reveal the tenuousness of entanglements and the need to promote a spatial hermeneutic that affords the capacity, following Miller, to understand the quality of associations and the relations that the world, including its human actors, articulate.

As the responses at the public meeting illustrate, the disturbance caused by the movement of the nitrates and the accompanying rhetoric may attempt to foster a relational habit of thought and resulting action but the spatial perception needed for such action is not readily practiced. In his comments concerning new materialism, Bergthaller (2014) acknowledges that "the playful expansiveness is one of the compelling features of new materialism, as it brings into view a range of interdependencies that are crucial if we wish to understand the contemporary environmental crisis" (41). While Bergthaller acknowledges the value of this expansiveness, he also identifies it as a drawback in that the approach focuses too extensively on the additive and inclusionary qualities of how matter comes to mean, neglecting the exclusionary tendencies that are also at work. The expansiveness of new materialism "leaves us with a theoretical and practical conundrum unless we also articulate a set of rules that would make it possible to draw distinctions between particular assemblages and different types of assemblages and to specify the conditions under which they can persist through time" (42). In other words, new materialism must account for the heterotopic quality of relationality and specifically the differing qualities of relations that exists across and within configurations. "What if there are two or more networks? How then to articulate the difference between associations *within* and *between* [emphasis in original] networks and—more important still—might it be the case that different networks hang together in different ways, are there different kinds of associations?" (Mol 2002, 70–71). Moreover, how might relations among the disparate be generated and how might we account for the quality of these relations? In Florida, for instance, concerns over the presence of waste find industrial effluent from power plants, nitrates from OSTDS, and phosphates from agricultural production becoming juxtaposed.

These sites of contamination all occur within distinct networks and places, whether that be the warm-water refuges created by industry, the warm-water refuges of the state's natural springs, or the inland lakes and coastal estuaries and beaches being tainted by algae. Yet the use of OSTDS by residents of Deltona has come to be juxtaposed with not only the dirtied water of the coasts but also the impending demise of the industrial warm-water refuges. Establishing relationality and subsequent responsibility is not a novel practice for environmentalism, including concerns as to waste's production and its resulting effects. Tracking the movement of waste as it moves downstream or downwind from its origin to the places in which it settles and disrupts has long informed the efforts of environmental advocates (Peeples 2011b). And the manner in which consumers are associated with the detrimental effects of industrial production has, as I have earlier discussed, also played a central role in informing environmentalism's relation to waste. The relationality that is fostered by and through the contemporary contaminated world encompasses, however, more than just these direct lines of causation, associations that are most commonly traced to capitalist production.

Contemporary relationality, and as a result response-ability to waste, must now be seen in terms of happenstance and the bringing together of unexpected and previously disparate networks, spaces, and nonhuman actors. If a potent rhetoric and mode of influence are currently lacking to move publics toward an understanding of the eventual demise of the industrial effluent that provides warm water to the manatees and to alert publics to the need to remediate the dirtied water of the springs so they will provide the needed refuge, then finding ways to link these spaces with the more potent rhetoric that has come to constitute the algae crises in other Florida waters may prove to be a viable practice. But as the persistent discourse of sustainability used at the MVC, the uncertainty of the manatees' movements, the resistance conveyed by some residents of Deltona, the lack of available funding, and the prominence given to apocalyptic images of algae blooms illustrate, articulating the link among these spaces and practices and prompting some response-ability from the public toward acts of ecological care entail much more than establishing some downstream connection. The current configurations that constitute our contaminated world thus present significant challenges as to how to articulate understanding of our emplacement within such heterotopia and how such understanding may inform collective action so that all may participate and share responsibility in ethically dwelling within this evolving world. And that challenge becomes further pronounced given the constraints placed upon human agency by these novel configurations and the uncertainty that emerges from contemporary entanglements.

THE PRESENCE OF BLUE-GREEN ALGAE

Located along the center of Florida's east coast, the Indian River Lagoon (IRL) stretches approximately 150 miles from Ponce de Leon Inlet on its northern end to the southern border of Martin County. The IRL is defined as an estuarine system, which "consist(s) of deepwater tidal habitats and adjacent tidal wetlands that are usually semi-enclosed by land but have open, partly obstructed, or sporadic access to the open ocean, and in which ocean water is at least occasionally diluted by freshwater runoff from the land" (Cowardin et al. 1979, para. 1). Like many other bodies of water in Florida, the IRL is entangled within various practices, with tourism, commercial fishing, and recreation generating important economic meanings of the place. "As of 2014, the annual economic value of the lagoon was estimated to be $7.6 billion, which included nearly 72,000 jobs and recreational opportunities for more than 7.4 million visitors per year" (SJRWMD 2020, para. 1). The economic reality of the IRL is shaped not just by various human practices but also through the range of nonhuman objects populating the waters. "Because the IRL comprises a transition zone between temperate and subtropical biomes, the IRL is considered a regional-scale ecotone and one of the most species-diverse estuaries in North America" (Lapointe et al. 2015, 82). The Florida manatee is just one of the diverse species entangled within this space, due in part to the seagrass, a vital part of the manatee's nutrition, that grows within the estuary.

Similar to Volusia–Blue Spring, the IRL has also become disturbed by abnormal levels of nitrogen. The increased levels have been attributed to numerous factors, including the migration of nitrates from the numerous residential septic systems placed within Volusia and Brevard Counties, counties that abut the northern edge of the IRL (Lapointe et al. 2015, 84). But the disturbance to these Florida waters and practices occurs not just through the movement of waste from septic systems but also, particularly in the southern half of the IRL, from agricultural practices. While citrus and produce farms are emplaced within this configuration, "the dominant land use is pasture and rangeland for beef cow-calf production" (Capece et al. 2007, 20). Florida ranks twelfth in the U.S. in beef production, with the majority of this practice taking place in "south-central Florida, largely on privately owned lands, and geographically overlapping some of the most sensitive wetland systems in the state (Swain et al. 2007, 2). The Florida cattle industry relies upon open-field grazing as the primary means of sustaining the beef supply, and meeting increasing consumer demand with no increase in the acreage of available grazing land has led to environmentally detrimental practices. Fertilizers containing phosphorous have historically been used to ensure an ample supply of grass on the cow pastures. Waste from this practice, particularly in the form of phosphorous, has moved from these grasslands through the Kissimmee watershed and into Lake Okeechobee (Capece et al. 2007, 20) resulting in consistent algae blooms and eutrophication of the lake. Land management practices by the agricultural industry have since drawn significant attention. As a result, best management practices have been developed to reduce the load of phosphorous currently entering the watershed and the lake.

Yet the occurrence of the algae blooms on Lake Okeechobee and the IRL results from more than just the practices of cattle ranchers. Inland water that moves to the IRL has been reconfigured and redirected, a product of decades of human efforts to manage and master the Florida landscape, starting with the 1916 Drainage Acts of Florida, which "permitted the creation of canals to drain uplands for agriculture, reduce flooding, and control mosquitos" (Lapointe et al. 2015, 83). Whereas inland water once moved through a series of streams and wetlands prior to entering the IRL, water that currently enters the IRL at the southern end from the St. Lucie estuary now moves first through Lake Okeechobee (83). A historically shallow lake whose southern end once consisted of marshes that filtered water south into the Everglades, Lake Okeechobee has been reengineered through a series of levees that constitute its current banks, allowing the lake to now hold a higher volume of water. Moreover, water leaving the lake no longer moves directly into the Everglades but rather in the direction of the urban and agricultural centers to the lake's immediate south and east (Havens and Gawlik 2005). Summer rains and unusual precipitation events such as hurricanes and tropical storms often stress the levees. To prevent water from breaching these barriers, canal systems have been constructed to relieve the strain on the levees and to move the excess water through the St. Lucie River and into the IRL on the east coast and through the Caloosahatchee River and into the Gulf of Mexico on the west coast. As the lake water flows through these rivers, so too do the nitrates, and phosphorous and the algae that feeds upon these discharged nutrients.

Legacy pollution is not necessarily a new phenomenon, given the history of manufacturing facilities and extractive industries that have either gone out of business or have mined all the resources from a particular location and have left contaminated soils and waterways after the industry has been shuttered. The legacy of the agricultural industry's increasing reliance on fertilizer is particularly active in the central and southern parts of Florida, including Lake Okeechobee and the surrounding waters. "One of the most vexing challenges posed by phosphorus is that it forms a strong chemical bond with soil and only a small portion is released at any given time, usually with heavy rainfall. It could be decades or even centuries before all the phosphorus in the agricultural basin washes into Lake Okeechobee" (Stern and Kornfeld 2020, para. 12). In addition to residing within the soils of central Florida, phosphorous also heavily populates the sediment of Lake Okeechobee. "The bottom of the lake, which was sandy for thousands of years, is blanketed with millions of tons of black muck containing an estimated 50,000 metric tons of phosphorus. Nobody knows how to remove or neutralize it" (para. 13). Similar to installing new OSTDS technology, motivating farmers to adopt best practices may impact the current levels of phosphorus that migrate into the state's waters, but the effectiveness of these efforts to purify and rid Florida waters of the agricultural waste matter may not fully materialize given the decades of accumulated waste that rests within the sediment of the lake and throughout the soils in the central part of the state (Swain et al. 2007, 10). As a result, the waste matter perennially circulating within Florida waters offers an opportunity to identify that the practices and perceptions that advance the human capacity to restore Eden to its pristine state are contrary to the configurations that constitute the contemporary dirtied world. In other words, humans may invoke practices to limit the current rate of toxic releases, but those efforts will not address the already existing contaminants that circulate among and shape contemporary configurations. In the following section of this chapter, I want to more fully discuss these limits on human action and locate the prospects for perceiving and dwelling among spaces of restoration more fitting to contemporary configurations.

THE FUTILITY OF CHASING THE MODERN, EDENIC MYTH

The need for a new perceptual apparatus that more fully accounts for the continued presence and agency of waste and that may more appropriately convey the prospects for restoration in a dirtied world emerges through the Everglades Agricultural Area Storage Reservoir Project, a proposed project intended to redirect the movement of dirtied water from Lake Okeechobee southward through the Everglades and away from the east and west coasts of Florida. As described by the South Florida Water Management District (SFWMD), the reservoir is anticipated to hold 240,000 acre-feet of water and include a new constructed treatment on a 6,500-acre wetland known as a Stormwater Treatment Area (STA) (SFWMD n.d., "Progress Continues" para. 1). In language that mirrors descriptions of the drain field that accompanies residential septic systems, the STA serves as a holding pond where the nutrient-loaded water from Lake Okeechobee can be treated, cleansed, and eventually released into the Everglades. The Storage Reservoir Project is the latest in a series of recent transformations to the Everglades landscape to address the existence of contaminated water. These STAs are located within three water conservation areas, which "are meant to act as a buffer between the Everglades Agricultural Area (EAA) and Everglades National Park to protect it from high phosphorus inputs" (Schade-Poole and Möller 2016, 6). The conservation areas not only include the STAs (which are constructed wetlands) but also flow equalization basins (FEB), which are "reservoirs designed to temporarily capture and hold stormwater runoff for release to the Everglades stormwater treatment areas" (SFWMD n.d., "Harmful Nutrients," 1). Within these STAs "non-native plants with an appetite for high levels of nutrients are being used selectively to help remove excess nutrients, so that native plants can once again thrive," and these constructed wetlands have been determined to be more effective than chemical processes in addressing the excess phosphorous levels (SFWMD n.d., "Water Quality Improvement," para. 1). When seen in relation to the existing STAs and FEBs, the proposed project to accommodate the flow of algae and phosphorous from Lake Okeechobee becomes one additional element of what can be considered as Florida's drain field. If an individual resident's septic system relies on natural processes and a specific bordered space by which to treat the waste prior to the dispersal of the treated effluent, then Florida's construction of the specific spaces abutting Everglades National Park can be seen as attempting to achieve the same ends, with the help of some non-native agents, however.

The newly constructed STA south of Lake Okeechobee is only one of the additional storage areas planned under the Everglades Agricultural Area Storage Reservoir Project.

> Without additional water storage and treatment, . . . ecological conditions and functions in estuaries on the east and west coasts of Florida will continue to experience adverse impacts due to excessive damaging regulatory releases from Lake Okeechobee during wet years, while the Greater Everglades requires additional flow with the proper timing and distribution to improve ecological conditions. (SFWMD 2018, ES-2)

As the comments by the state attest and which are echoed by numerous other organizations, the expansion of the treatment areas will address some of the incursion of waste in the coastal estuaries and protect those ecosystems; these new ecologies will also beneficially move water toward the Everglades. The Audubon Florida (2016) organization, for instance, notes that the treated water moving into the Everglades from the STA can help mitigate the "massive seagrass die-off taking place in Florida Bay occurring because of insufficient freshwater" (1).

Yet understandings of human agency in constituting such restored space and the efficacy of human actions to purify and restore the tainted water and Florida landscape must be considered in relation to the uncertainty and unpredictably resulting from vibrant matter's capacity to push back against these human intentions. Nonetheless, state agencies draw from the familiar rhetoric of balance and restoration to describe the efficacy of these drain fields and created spaces. For example, the SF-

WMD describes that the "STAs provide another bonus—prime home and visiting territory to wildlife including wading birds, ducks and American alligators. A variety of nature-based recreational activities are allowed at several of these wetland locations" (SFWMD n.d., "STA 5/6," para. 2). The recreational benefits of the STAs are also referenced by the Fish & Wildlife Foundation of Florida (2015): "This trail hotspot is one of the best birding and wildlife viewing locations in Florida. Constructed to filter agricultural runoff from water destined for the Everglades, this complex of four water impoundment cells is an impressive refuge for birds and a mecca for birders and photographers all year long" (para. 1). Like the MVC in which manatees can be encountered in a dirty space, the STA is also a site in which the presence of waste exists alongside human activities and encounters with wildlife. In another instance of contemporary incongruity and heterotopia, the STA juxtaposes recreational activities with waste treatment. The entanglement of these disparate practices does not suggest, however, that their incongruity is reconciled, even as the state discourse that relies on the familiar rhetoric and perceptions of restoration suggests. What fosters the potential to interpret and respond to such a space in ways that move beyond the familiar figure and perception of restoration is the recognized vibrancy of the emplaced dirty matter. Not only does waste remain in the space; it also challenges notions of human mastery and continually calls upon the human actor to creatively dwell within such a space of contamination. The STAs, as exemplars of contemporary heterotopia, are not spaces devoid of dirt and cleansed by human actions. Nor are they dystopian spaces, of material invocations of mourning for the lost Eden and unspoiled wonder that existed prior to human incursions. Instead, these are spaces that illustrate the emergence of contemporary tensions and the resulting acceptance of a human vulnerability needed to effectively engage in the continual negotiations between others entangled in such configurations.

For example, the process of removing the nitrates and phosphorous in the STA's relies upon the actions of various plants that serve as natural filters of the pollutants.

> The STAs are designed to operate so that aquatic plants, with associated microbial and biogeochemical dynamics, are used to remove phosphorus from the water. The runoff is directed to flow through different treatment cells made up of a variety of aquatic plants. These plants remove phosphorus by absorption for use in metabolic processes, and also store it in plant material and cycling plant detritus. The process by which it this [sic] occurs is that first, the water flows through treatment cells that contain emergent aquatic vegetation such as cattails." (Schade-Poole & Möller 2016, 12)

Cattails, vital to remediating the tainted water, however, present a challenge to other species within the Everglades. "The native sawgrass populations are adapted to live in a low nutrient environment. With increasing phosphorus within the Everglades, plants that favor a more nutrient rich environment such as cattails, are dominating the marsh and slough communities" (9). Because of the use of cattails and other aquatic plants that can filter the phosphorous, the STA's take on an ecology distinct from that of the Everglades. A central concern is then to ensure that these species, along with the phosphorus, do not readily migrate and alter more of the landscape. But even with human design and oversight, the stability of the configuration, as measured in its propensity to rid the water of waste, is also dependent on the uncertain actions of the other-than-human.

> During times of drought and decreased rainfall some areas within the STAs dry out, therefore efficiency and performance of STAs are negatively impacted resulting in high TP [total phosphorous] concentration in the outflow. Other extreme weather events with high winds and heavy rains increase water flow and erosion, which can result in the mixing of phosphorus rich suspended sediment into the water column, thus increasing TP concentration in water flow. Additionally, hurricanes and storms can damage and uproot vegetation communities in STAs and decrease the effectiveness of phosphorus removal with these aquatic plant communities. (14)

Rather than relying on a rhetoric of restoration, purity, balance, and human primacy, any public discursive practice affiliated with the configuration of the STA's must account for the configuration's variability and potential to be altered by unpredictable, other-than-human actors. In this regard, the human subject emerges not as an agent capable of subduing environmental disturbances and restoring some Edenic order but rather as a performer among many others whose abilities are constrained and enabled by the particular configuration.

The limits to restoring a clean and pure landscape and of generating sustainable ecologies become further evident when considering the lack of available space that can be reconstituted to address the voluminous presence of phosphorous that exists within the soils surrounding Lake Okeechobee and that are embedded within the lake sediments. Simply put, the state cannot construct a drain field with enough storage capacity to hold and treat the volume of tainted water.

As the persistence of the dirt and the subsequent maintenance and questionable performance of the STA's in Florida illustrate, the human capacity to conquer and arrange the world to satisfy its intentions has become increasingly challenged. Whereas mastery and domination once marked the movement across the nation, the posthuman tourist traveling through the contemporary Florida landscape encounters uncertainty and constraint. The heterotopia that mark posthuman America are thus not just spaces of multiplicity in which waste and human recreational activities are juxtaposed. These are also spaces of conceptual incongruity, configurations fostering a vulnerable, posthuman subject whose relation with contaminated matter becomes shaped both by the possibilities afforded by human creativity but also by the constraints emplaced by unpredictable and malevolent nonhuman actors.

The Florida STA's, as noted, include various non-native species who enable the processing of the excessive waste material, but these participants in the space are far from the obedient dog submitting to the command of a human master to stay put. Actions by nonhuman actors to disturb and push back against human intentions to reconstruct and purify the landscape have become evident in other STA's. For example, the Lakeside Ranch Storm Treatment area, located on the northeast side of Lake Okeechobee and designed to manage excessive phosphorous prior to its flow into the lake, has faced significant and unpredicted disturbances and rearrangements.

> In addition to low water levels, wildlife interactions have negatively impacted vegetation enhancement activities, particularly the establishment of desirable plant species. Heavy caterpillar populations coupled with hog damage have deteriorated the conditions within the cells. Fish kills from the decreasing water levels have also resulted in an increase in avian activity adjacent to the outflow sampling stations. The increase in wildlife disturbances and the significant loss in vegetation has most likely contributed to the recent poor performance in the STA. (SFWMD n.d., "Lakeside Ranch," 1)

The challenges to attaining the desired human ends of restoration and balance intended by these STA's clearly assert that the lofty human ambitions are ill-suited to the complex entanglements of contemporary heterotopia and make evident that uncertainty generated through the antics of various other-than-human actors, including and beyond waste matter, is a primary trait of a dirty world. Uncertainty may have informed Beck's ([1986] 2011) risk society and his anticipation of a reflexive and more politically engaged public. In many respects, however, Beck's hope for a politically awakened subject more attuned to risk has failed to materialize. And while risks remain, uncertainty has evolved given the increased presence of and surprising juxtapositions involving vibrant, dirty matter. As the unknown movement of manatees and the unpredictable disturbances to the STA's illustrate, uncertainty is no longer strictly associated with questions of whether risk exists or the authority granted to those processes and knowledge by which risks are defined. Ecological care is no longer constrained to reducing the vulnerability and risk posed by the dangers of toxins and industrial waste. Vulnerability now encompasses an extended body of uncertainty, a contemporary ecological ambiguity aligned with the constraints and possibilities afforded by a world populated with vibrant dirty materials and other unexpected actors forging unpredictable and dissonant confederacies that provoke humans to constitute new environmental relations and ethics.

PUBLIC PARTICIPATION AND THE MISCREANTS

One of the attractive promises of new materialism and of rhetorical theories used to interpret space is that reconfigurations brought about either by intentional human efforts or by happenstance and unpredictable nonhuman agents may lead to novel and more ethical practices and ways of being in the world. The various spaces discussed in this chapter, whether they be the paradoxical site of Big Bend or the reconstitution of space such as the STA's that attempt to address Florida's water crisis, exhibit a novel and heterotopic arrangement, but their potential to invoke an enhanced environmental ethic can be questioned, particularly if one measures ethics in relation to familiar frames and perceptions. As I have earlier mentioned, one of the central considerations of this book is that familiar habits of thoughts and norms are no longer appropriate to a world that has become contaminated, and in a world of prevalent dirt, conventional practices associated with environmentalism are no longer viable. Subsequently, inquiry must attend to identifying new hermeneutic practices that shape understandings of this dirty reality, including the figurations that better reveal our prospects for living with the dirty matter and the normative concerns that such configurations provoke. The material-discursive rhetoric that constitutes Big Bend as a site of sustainability can be critiqued as corporate greenwashing if one draws from familiar habits of critique, yet the heterotopic quality of the site and its entanglement across dispersed sites and practices reveal that the space needs to be interpreted as more than a tired attempt to generate a corporate environmental identity or as another example of hegemonic, institutional control over a public's environmental imaginary. Carr and Milstein may associate public indifference to the heterotopia of Big Bend with willful ignorance, yet decades of failed practices to keep the world clean along with false prophesies of a coming dystopian planet make environmentalism also complicit in generating a public that has come to accept the presence of dirty matter. Drawing attention to the environmental harm resulting from carbon-based energy production may promote some awareness and hopeful action to cease some of the dirty emissions from the power plant, but such familiar and increasingly ineffectual critiques fail to account for the dirtied reality that presently exists and the means by which humans perceive and act within such dirtied spaces. I do not mean to suggest that we should neglect the harm that carbon-based energy inflicts upon the planet nor cease efforts to reduce or eliminate it. Such efforts may limit future amounts of dirt, but they do nothing to address our capacity to currently live within the already dirtied spaces of the planet. Subsequently, our perception of heterotopic spaces such as Big Bend needs to expand beyond immediate concerns of eliminating the waste. Moreover, as in the case of Big Bend, the eventual dismantling of the carbon economy and its dirtiness means that we must attend to the uncertainties that characterize the new ecology manatees will find themselves emplaced. The familiar ecologically ethical path to follow in this scenario is to advocate for the restoration of the warm-water springs; yet the uncertainty of the manatee's movements must be a part of any effort to influence public perception and action.

As a result, caution must be taken to quickly cast aside and to criticize the prospect of dwelling with waste as indifferent to environmental care. Given the prevalence and persistence of waste and the increasing uncertainty of human actions to remove it, we have no choice but to find ways to live with it. The critical task, therefore, is to better emplace the human, as a creative and vibrant force but one also with an acknowledged vulnerability engendered by the other, within the contradictory conceptual and material heterotopia of the posthuman, contaminated world. As the continued presence of dirtied water in Florida shows, additional technologies and practices are being developed as a response to existing challenges and to the prevailing uncertainties. Obviously, attention must be directed to understanding the processes involved in the emergence and implementation of these efforts not through familiar frames and discourses but rather through new interpretive lenses and figures that better reveal the appropriateness of such novel practices in an entangled and uncertain contaminated world. While some may lament the position granted to science in relation to environmental crises, dwelling in a contaminated world will require an increasing reliance on this specialized process of research and implementation.

While the role of techno-science in a contaminated world deserves serious consideration, the dirtied spaces of Florida described in this chapter reveal an additional line of inquiry concerning the way the relationality and uncertainty that constitute contemporary configurations intersect with the ethics of the publics' responsibility. As I mentioned in my opening discussion of Big Bend, the USFWS recently reclassified the manatee from endangered to threatened. This reclassification, while seemingly indicating the success of past conservation efforts, was not met with universal acceptance, however, with some groups, including Save the Manatees (STM) suggesting that the move is premature. In its opposition to the reclassification, STM generated a document that characterizes a relational mode of thinking and rhetoric in which it articulates associations among the imminent loss of the industrial warm-water effluent, the degradation of the warm water springs because of the excessive nitrates, the loss of the vital food source of seagrass in the Indian River Lagoon due to excessive algae growth, and the lack of available fresh water in the Florida Bay (Tripp 2016). The intention of STM was to illustrate that the manatee, a mammal whose survival depends upon its movement to adequate feeding sources and warm-water refuges, will face significant population challenges given that it, like its human co-inhabitants of these places, finds itself dwelling in spaces of dirt. Citing information from the now defunct Warm Water Task Force, Tripp (2016), writing for the STM, notes:

> Given that the majority of the Florida manatee population now use power plant outfalls and natural springs for warm water, the future status of the Florida manatee will be affected by what happens to these sites. . . . The protection and enhancement of natural sites and how we address the loss of industrial sites will set the course for the manatee's future. (4)

The STM sees the public as actively associated with addressing the uncertainty tied to the manatees' movements from the industrial effluent to the natural spring water.

> For example, if the Service requires the power companies to fund some of the work needed to research and plan for manatees' transition away from artificial warm water, perhaps by adding a fee per kilowatt hour that would be paid by consumers, it would be more likely to sway the PSC and the public towards a supportive position for an endangered species with an understanding that these actions would help the species move towards downlisting and recovery. (5–6)

To counter the claims in the STM letter, the USFWS (2017) contends, through reference to scientific models, that the various dirtied spaces that constitute the manatees' habitat would not result, at least in the next fifty years, in a significant population decline. While the letter's failure to move the government merits attention in terms of its rhetorical efficacy, the letter's suggestions to initiate public engagement to address the ecological disturbances speak to the need to attend to an ethic of relationality and the quality of associations that link the disparate and dispersed spaces, practices, and actors within the contemporary dirty world.

As the STM letter advocates, publics will be asked to share a financial burden to help address the uncertainty that exists with respect to the manatees' movements once the power plants are decommissioned. This call for public financial participation also takes place within the rhetoric in which the state of Florida has attempted to motivate public action to mitigate the nitrates in the warm-water springs through an additional public financial burden. Publics would thus be tasked with funding efforts to restore the needed warm-water springs while also funding the research to determine the probable movement of the manatees from their current industrial warm-water refuges. Public participation and ecological care is thus tied to economics. Given the resistance shown by some residents of central Florida to the incentive program to alter the current OSTDS, the additional financial burden on publics to support research to resolve the uncertainties of the manatees' movements would also likely face significant pushback. Given the publics' resistance, it may be precariousness, specifically as propagated by those miscreant vibrant actors and their unpredictable ability to disrupt human efforts, that may be the most potent rhetoric to address the tainted springs and the uncertainty of the industrial warm-water refuges.

Disturbances to the STA's as described earlier along with the limited, available space for the state to construct the needed drain fields mean that the dirty water that has moved to the valued Florida coast lines and estuaries and that has prompted significant media attention will persist and not easily go away. Mitigation efforts to address the blue-green algae, therefore, will need to continue, including those aimed at addressing the excessive discharge of nitrates from OSTDS that may flow into Lake Okeechobee and the IRL. Subsequently, those vibrant miscreants responsible for disturbing the performance of the STA's are also, by way of relationality, a key rhetorical actor involved in moving the state and publics toward the remediation of the warm-water springs. If the residents of Deltona were linked to the remediation of the warm-water springs due to the potency of the material-discursive rhetoric aligned with the dirtied coastal waters and estuaries, then the continuing uncertainty generated by the troublesome disturbances of varied nonhuman actors will continue to articulate the emplacement and actions of these publics. As I mentioned in my earlier discussion, residents of Deltona have been resistant to altering their OSTDS technology based on their assertions that the processes they use to treat human waste are not related to the blue-green algae crisis that has generated so much attention and concern. Geography bears this out, since the waste from the residents of Deltona will not leach into the IRL or into Lake Okeechobee given that this population is emplaced within the St. John's River Watershed. Their effluent flows north as opposed to east to the IRL or south toward Lake Okeechobee. In this ironic instance, traditional lines of cause and effect reasoning disrupt the contemporary articulation to sponsor a wider relationality. Yet emplacing publics within this geographically dispersed configuration may indeed prompt the needed ecological care. The resistance to the incentive program by way of the residents' disarticulation must, therefore, be considered in relation to the public's lack of engagement to address the nitrates in the warm-water springs. Given that Volusia–Blue Spring is a part of the St. John's River Watershed, the effluent from the residents of Deltona inarguably impacts this warm-water refuge. Their collective waste clearly contributes to the degradation of the spring and to the needed manatee habitat, yet this linkage has not generated the degree of public attention and movement garnered by the incursion of blue-green algae.

The efforts to articulate residents of Deltona with other dirtied Florida waters thus raise ethical concerns as to the appropriateness and the political efficacy of such material-discursive arrangements. If motivating publics to engage in addressing the dirtied spring waters falls short by emplacing residents within the bordered watershed that constitutes the spring, then consideration must be given to reconstitute the space in which they are entangled. In this regard, how relations become articulated and how spaces become constituted become major concerns for a contemporary environmental ethic. Connectivity and relationality have consistently informed environmental thought, as evident in the effectiveness granted to the "blue planet" image and the various downstream and downwind approaches that have productively expanded understandings of industrial production, environmental justice, and environmental degradation. But connectivity and relationality in the posthuman world require a reconsideration of the way associations develop and hang together, especially when posthuman actors can reconstitute connectivity and associations beyond the traditional, humanistic perceptions of cause and effect and in ways that promote novel discursive-material arrangements. For example, to what extent are the actions of residents of Deltona, specifically those resistant to remediating their septic systems, associated with the other-than-human actors disturbing the performance of the STA's and the subsequent efforts to remediate the dirtied waters? How are each of these actors implicated and subsequently articulated within the configuration that includes warm-water springs, manatees, coastal real estate, and Florida's reliance on tourism? In this chapter, I wanted to articulate the associations among various geographic spaces of Florida to provide some initial consideration of the challenges brought about by contemporary, and seemingly disparate, entanglements. As my discussion shows, an encounter with waste and with novel configurations does not necessarily move human subjects toward an enhanced ethic of care toward the nonhuman other. The challenges manatees face in dwelling within the dirtied Florida waters and the uncertainty brought about by these configurations have failed to provoke some transformative human response. Such immobility may be due not to willful ignorance but rather to the persistence of an untimely perceptual apparatus and corresponding environmental ethic, as in the form of sustainability and remediation, and the challenges to appropriately figure human emplacement within the disorder of uncertainty and within novel and seemingly dissonant relations. What the configuration that constitutes Florida's dirtied waters reveals is that contemporary ecological care cannot disregard the uncertainties provoked by the novel configurations that constitute this dirty epoch and must embrace the requisite vulnerability and response-ability to ethically engage in dispersed and dissonant relations with various and surprising others.

4

✢

Reorientations to Risk

In 2003, the United States government established the Office of Legacy Management (OLM), a branch of the Department of Energy (DOE), that took charge of remediating and monitoring various locations contaminated during the production of the nation's nuclear weapons. As of 2021, LM was responsible for remediating and monitoring over 100 of these sites, including abandoned uranium mines, decommissioned uranium processing plants and nuclear weapons production facilities, and former nuclear test sites. A number of these locations have been designated for re-use and are now wildlife conservation and habitat development areas, cultural heritage sites, and acreage that provides western ranchers with agricultural and grazing space (OLM n.d.). While scholarly attention has been directed at these remediated nuclear weapons production facilities, this research has largely focused on bringing to light the way these sites contribute to an often uncritical remembrance of the country's nuclear heritage. The research has followed an important but common thread in memory and heritage scholarship noting how the symbolic and material resources of these remediated places advance a narrative of the importance of nuclear weapons production to ensuring the security of the nation but fail to account for the human, ethical, and environmental effects of such practices (Kirk 2012; Krupar 2007; Krupar and DePoe 2007; Taylor, 2010; Endres 2018). The critiques advanced in this research mirror the larger challenge many scholars have extended to remediation, mainly that such practices promote the invisibility of toxins and subsequently reaffirm the industrial, military, and consumer practices responsible for generating waste. The OLM remediated site of

Weldon Spring, Missouri, however, offers a starting point to investigate the potential for additional effects and new forms of ecological care when waste is no longer absent or emplaced within the boundaries of marginalized communities.

Taking exit 9 south off I-64 in Missouri leads one to the Weldon Spring Conservation Area, roughly 8,400 acres of mixed forest, pasture, and wetlands bordered to the east by the Missouri River and to the north by middle-class suburbs. With numerous hiking trails, including connections to the statewide Katy Trail, fishing ponds and lakes, and access to the Missouri River, the conservation area provides residents of the populated St. Louis metropolitan area with an opportunity for familiar forms of recreation and isolation from nearby urban populated areas. The conservation area also offers a unique opportunity, something quite out-of-place to the traditional and more recognized activities and landscape features one would typically find in such spaces. The Weldon Spring Conservation Area is also home to the Weldon Spring Site, the previous home to a TNT and DNT production facility during World War II as well as the location for uranium processing during the 1960s. After the DOE no longer used the site for weapons production, the location, as with many military-industrial locations of the Cold War, was littered with contamination. Rather than removing the radioactive and industrial contaminants and storing the matter off-site, OLM constructed a containment cell to permanently house the toxins on the Weldon Spring Site. The containment cell is publicly accessible and a vital part of the public heritage and recreational experience at Weldon Spring, a location that not only affords a bodily encounter with a nuclear landfill but that also situates a radioactive landfill within surrounding recreational and residential areas. Such a spatial configuration has led officials from OLM to contend that Weldon Spring affords an opportunity to enhance the organization's level of public trust. As OLM officials contend, the public access to waste promotes better understanding of the practices needed to ensure the toxins are stable and controlled: "'If you put up a fence, all that communicates is fear,' said Pam Thompson, the Weldon Spring project manager. 'The only way to defeat fear is knowledge'" (Simon 2002). The heterotopic space can thus be seen as an effort to constitute a posthuman subject, an ecological identity crafted not through the absence of toxins but rather through the capacity of publics to encounter and be present with the vibrant matter of waste.

In this chapter, I want to explore the potential of such a transparent encounter with waste by identifying how spaces such as Weldon Spring and other sites of restoration where waste is a visible and present object operate not as utopias that promote the modern prowess of human rationality and science nor as dystopias that convey the effects of the uncontrolled excesses of modern practices; instead, I propose that sites that promote encounters with dirty matter work as other-spaces, as heterotopia that are located in some in-between place and that exhibit some hybridity between the familiar figures of utopia and dystopia. It is in the heterotopic juxtaposition and subsequent tension of the promise of utopia and the failure exhibited in the dystopian, of the uncertainty and certainty, and of the beautiful and ugly found in Weldon Spring and similar spaces where waste is encountered that generate their contemporary meaningfulness and offer an alternative perception that may better suit the tasks of dwelling in dirt. Encounters with waste are not influential, transformative events that move publics toward using less and advocating for stricter emission policies. Rather, the incongruities that characterize these spaces make possible a reorientation to the responsibilities one has when living in a contaminated world. Waste disposal has typically been cast as a practice outside the realm of publics, evident through the quotidian practices of placing one's garbage curbside and in those specialized institutions that manage and secure waste objects and transport the dirty matter away from publics. Public encounters with waste, specifically those involving sites of recreation, invoke publics to engage in the contraries and incongruities of the contaminated world, which in turn call upon some perceptual apparatus by which to reconcile the disorder and imbalance. It is the capacity of these spaces to generate this disorder and the perceptual resources that may inform efforts to dwell with dirt that I wish to focus upon in this chapter.

To reveal the characteristics of such spaces of heterotopia, I want to situate Weldon Spring within recent alternative spatial and representational practices, specifically Land Art and The New Topographics, aesthetics that have become associated with waste and with spatial relations between the human and the nonhuman. An art movement that arose during the latter decades of the twentieth century, Land Art is most often associated in America with Robert Smithson. Frequently characterized as neo avant-garde, Land Art is characterized by the construction of "'environments,' 'installations' etc., which can be, but not necessarily always is, an expression of some environmental awareness in a modern ecological sense" (van den Berg 2006, 380). Smithson's *Spiral Jetty*, a spiral-shaped human-construction of mud, salt, and rocks that extends out into the Great Salt Lake has become a frequently referenced example. While some Land-Art may indeed attempt to generate awareness of a specific environmental problem, the movement as a whole can be seen as attempting to establish a renewed understanding of the relation of humans and matter.

> These artists tried and try to create and develop an art which is not only organic in itself, for instance, by using organic materials, but also try to adopt and include principles of the *natura naturans*, by including natural transformation, growth, decay and the like as calculated elements in their works. In this respect the application of natural materials in the open leads to the paradoxical constellation that in pursuit of a more sustainable relationship with nature the sustainability (or should one say: durability?) of their "real" works is frequently quite limited. (380–381)

Through the capacity of various actors and practices to modify the composition, the work of art moves from a permanent and stable act of the human creator to a continuously evolving and uncertain form and aesthetic produced by the unpredictable movements of a confederacy. The spatial composition as art is no longer about permanence but rather about the unexpected alterity and aesthetic that emerge from such unpredicted change.

While The New Topographics movement is commonly associated with the 1970s and the collection of photographers who sought to alter the conventions of landscape photography that privilege pristine, natural spaces, recent affiliations with the movement have sought to represent the way waste and toxins have become a primary feature of the American landscape. Peeples (2011a) explores the work of one of these photographers, Edward Burtynsky, and claims that such images of "beauty in destruction--stunning images of devastated landscapes" (376) generate an experience of the toxic sublime, a response that "encourages contemplation of the viewer's position within a polluted world" (375). By making waste visible and by juxtaposing the incongruity of the beautiful and the destructive, these particular images can be seen as heterotopic, disturbing prevailing perceptions of the utopian landscape while also challenging prevailing perceptions of waste's dirtiness. Following the familiar line of critique, Peeples contends that the alterity of the photographic composition prompts a response toward greater environmental awareness. Plainly stated, this particular aesthetic is said to prompt the familiar effects of guilt and remorse and the resulting desire to prevent such environmental destruction from occurring elsewhere. Such a response may offer some hopeful promise to save and preserve other spaces, but these common affects tend to exclusively regard these dirty spaces as simply incentives to engage in preservation and protection, essentially disregarding the disturbed spaces after the initial perception has been provoked. The spatial-material encounters with configurations of waste I explore in this chapter offer an alternative hermeneutic, calling publics to acknowledge one's environmental responsibilities and enact a form of ecological care for these contemporary heterotopia.

WASTE AS A SUBLIME, POSTHUMAN TOURIST OBJECT

While a significant number of postindustrial contaminated areas, brownfields, and landfills are fenced-off and publicly inaccessible, others have been remediated into commercial developments, have taken the shape of what is often referred to as their "natural habitat," and have become municipal parks and regional recreational spaces. For many scholars, these accepted practices of remediation, nonetheless, serve to hide and make invisible the toxins and past effects of these dangerous industrial practices, illustrating the prominence attached to absence as a means for modern society to perceive and live with the existence of toxins. "Waste is an ironic testimony to a desire to forget. Landfills, in other words, make their appearance on and in the landscape as a material enactment of forgetting" (Hird 2013, 106). Instead of directing understanding to the practices that generated the need for remediation and waste processing, the remediated landscape aligns human relations to waste with those after-production processes that can reportedly maintain the desired distance and the needed safety from the dirty matter. Such distancing, as part of the overall practice of waste disposal, has contributed to the "tendency for individuals to experience waste in a fleeting, almost ethereal manner," which in turn "translates into a societal failure to truly experience and therefore acknowledge waste" (De Coverly et al. 2008, 291). Even with the ubiquity of waste in modern society, the practices that instill absence and forgetting have generated a perceptual distance with the object while instilling a reliance on and belief in the primacy of human rationality to maintain such social order. The ability to remove toxins from daily life and to control any negative effects that the waste may present has led, as some contend, to a public acceptance and even complacency toward the dirty matter. Pezzullo (2007), for example, notes that "Rather than extraordinary, toxins sometimes seem all too ordinary.... Because the label appears almost everywhere, it becomes difficult to discern when calling something or someone toxic is worth our attention and concern" (63). And as Peeples (2011a) echoes, "Because of the toxins' invisibility and banality, individuals often attend to environmental problems not because they are the most dire, pressing, or dangerous, but because they are the most evocatively articulated" (374). As a result, many contend that generating a critical subject and a revised rhetoric and relation to waste involves moving beyond the perceived and real absence of waste by placing subjects within an encounter with a present, discursive-material artifact that can "evocatively articulate" and generate the needed disturbance to shake such public complacency.

While tourism has long been associated with acts of leisure, the practice has recently become associated with political acts and as a means to promote a renewed environmental consciousness. Ecotourism is one clear example of an effort to promote attunements with the other-than human and to foster an ethic of ecological care. Toxic tours, often associated with the work of environmental justice advocates, are also designed to promote a political subject by immersing tourists within spaces of contamination. When situating visitors within contaminated environments, as Pezzullo (2007) explains, "the materiality of a place promises the opportunity to shape perceptions, bodies, and lives with respect to the people and places hosting the experience" (9). By becoming immersed in the spaces of contamination, tourists come to bodily identify with those individuals who regularly experience the psychological and physical effects of being entangled with toxins. An additional tourist practice of contemporary society that is suggested to disturb complacency is what scholars have labeled dark tourism, manifest in expeditions to places wrought by industrial disaster. More than just a distant and superficial encounter with the effects of a human-induced disaster, such tourist practices also have political ends, with Chernobyl standing as the definitive dark tourist location (Stone, 2013). The lure of such places rests, as Stone suggests drawing from Foucault, in their constitution as "heterotopias," and as spaces of "contradiction and duality" (80), manifest in the spaces' bewildering juxtaposition of the ordinary and conventional with the out of place. As Stone explains, the dissonance of Chernobyl emerges through an interpretive juxtaposition that unsettles conventional categories while presenting opportunities for new meanings to emerge. "It is where the dead zone of Chernobyl offers the strange mixing up of conventional notions of ruins and monuments" (87). This positioning of odd and traditional material and interpretative arrangements generates a duality within the subject. "It is the juxtapositions of the real and the familiar with the surreal and the alien that allow tourists to consume not only a sense of the ruinous beauty and bewilderment, but also a sense of anxiety and incomprehension in a prettified place that mirrors our own world" (87). The incongruity experienced within spaces of dark tourism thus allows the contemporary subject to move past the accepted practices of meaning making and interpretation that cast a particular order on our world and toward the incongruity, uncertainty, and disorder that better reveals our posthuman reality.

While incongruity and incomprehension may be characteristic of the posthuman experience, the formation of such a perceptual encounter has historical influences. Other-than-human objects have long been attributed to generating responses of incongruity and disruption with one of the more notable being the way various natural objects invoke the sublime. Initially, the sublime encounter was attributed to objects of natural creation with Yosemite's El Capitan and other wonders of the American

West serving as iconic representations of the experience. Yet as Nye (1994) argues, technical objects also are able to invoke the feelings of awe, wonder, and fear that characterize the sublime. As they were introduced in the 19th and early 20th centuries, objects such as the steam locomotive, bridges, and skyscrapers were able to affect individuals due to these technical objects' "massive scale and permanence" which "made them appear to be triumphs over the physical powers of nature" (56–57). The encounter with these objects and the resulting affective experience, according to Nye, were generated through an initial interpretive dissonance and disorder. An encounter with such objects prompted a "moment of intensity" (2), a "mental shock" (6) that led to "a temporary dislocation of sensibilities that forced the observer into mental action" (6). An encounter with these technological objects thus becomes rooted in an interpretive challenge, as subjects sought to appropriately classify such wondrous objects into existing categories and frames of reference or to invent novel forms of understanding that could account for relating to these strange materials.

The reconciliation of the perceptual disturbance and the emergence of an appropriate perceptual frame came about through the tenets of modernity, specifically with respect to the ideology attached to the supremacy and power of human reason. The sublime experience of encountering a locomotive thus emerges not just through the scale and potency of the technological machine but also through the appreciation that humans can engineer and build an object with such force. The affective stance that emerged from the early technological wonders was one of wonder, amazement, and "a celebration of the power of human reason" (Nye 1994, 60). But if the technological sublime were to foster a belief in the power of human reason to master, and moreover, to even better nature, the creation of the nuclear bomb served as a turning point for how humans would experience the technological sublime. Humanity could unequivocally exceed natural forces and processes as it now obtained and made real the absolute power of destruction. Technology could not only instill wonder and awe; it could also generate an unimaginable terror. "The rocket and the atomic bomb would reshape the real world" by "bringing terror back to the technological object and erasing any illusions that science was intrinsically beneficent" (225). The sublime experience that was prompted by technical objects and the interpretive challenge such objects posed thus became embedded in an emerging skepticism toward human reason, a distrust rooted not in the competency of human primacy but rather in an incertitude of the master's beneficence.

Subsequently, the technological sublime is always embedded in the juxtaposition and contraries of modernity where on one hand technological objects generate a sense of wonder and awe, with promises of utopian order, but also generate a sense of terror and apprehension over the objects' harmful and dystopian possibilities. This same duality has been recognized and attributed to toxins, one of the more prevalent objects of modern technology. As Pezzullo (2007) describes: "Precisely because they promise us both dreams (of science, technology, and upward mobility) and nightmares (of insignificance, fallibility, and illness), toxins appear as the extraordinary, as that which exceeds our everyday understandings, in the national imaginary" (60). Waste's capacity to generate such psychic and physical disturbance has been central to banishing the dirty matter to places of absence. Modern waste disposal practices are thus aimed at reconciling the incongruities of a modern subject's encounter with this dirty technological object through its proper placement. By making waste distant and absent, the ambivalence facing the posthuman subject is cast aside, leaving the subject to experience the promise of technology without the risk. But the ubiquity and agency of waste have made making the dirty object spatially absent more difficult. As a result, the posthuman subject must contend with the perceptual incongruity that contemporary configurations of dirty matter foster.

DEFAMILIARIZING WASTE'S ABSENCE THROUGH ART

While photography has established conventional practices for constituting the American landscape, it has also, following the ubiquitous presence of toxins, become an important means to reconfigure human relations with and understandings of waste. DeLuca and Demo (2000) argue that images, such as Carleton Watkins' late-nineteenth-century photographs of Yosemite, "are founding texts in the construction of a wilderness vision that has shaped the contours and trajectory of environmental politics" (243). These images brought to the public environmental imaginary the experience of the sublime and a particular mode of perception by which to interpret the landscape. Much like technological objects such as the locomotive invoked awe and wonder and an affect of bewilderment, so too did the encounter with these visual representations of the natural landscape. The sublime thus became attached to national spaces of natural incomprehension and served as a key rhetorical trope used to preserve these other-than-human objects and spaces. Emerging practitioners of environmental photography, however, devoted little attention to the debasement of the landscape, particularly those spaces becoming increasing polluted by industrial practices. If visual representations were to prompt awareness of the detrimental impacts generated by these practices and the need to address the nation's dirtied spaces, new modes of visual representation, practices that moved beyond the prevailing sublime renderings of the other-than-human landscape, needed to be developed.

Through the advent of alternative means of representation, photography has become a critical resource to advocate for the injustices experienced by those living in spaces where toxins are present. Scholars and environmental advocates contend that images of toxic landscapes can construct an alternative reality and may invoke political action through novel practices of representation. Photographs of individuals living with the effects of toxins, as Barnett (2015) explains, may mediate the distance between the marginalized communities who embody the effects of the presence of waste and those privileged communities whose economic resources afford the ability to make toxins absent. "Given that most people cannot—or will not—travel to polluted places, images play a vital role in the public's ability to apprehend the precariousness of lives lived in toxically assaulted communities" (409). The capacity to establish such affect in the viewer is generated by toxic portraits,

> close-up, in situ *photographs of people within toxically assaulted places in which the relationship between pollution and the precariousness of life is illustrated through a range of visible (industrial, environmental, and corporeal) referents, accompanied by verbal captions and descriptions, which goad the spectator to interpret the portrait through the subjunctive voice*" [emphasis in original]. (410)

The persuasive potency of these portraits rests in their capacity to reveal entanglements and to generate particular configurations of space: "the form of the toxic portrait depicts not only a person, but also a person who is located within a particular context or condition that is the consequence of certain political, economic, and social relations" (412). These portraits enact a similar ethic as that prompted by toxic tours. The subject's emplacement within such material and ideological relations provokes an ethic of humanity and the innate human desire "to eliminate certain forms of socially facilitated forms of living and dying that pervade contemporary societies" (422). Political action and social change occur not through the immersion of the viewer within the material configuration, as in the case of toxic tours, but rather in the affectability fostered by seeing other humans entangled with the hazards of modernity. The influence of these images thus rests on the affectability of the viewer to recognize and respond to the other.

Portraits that offer incongruous juxtapositions of the human body with various objects associated with toxins are not the only visual rhetoric practice used to constitute the prospects for dwelling with contamination. As mentioned, Peeples (2011a) draws attention to the toxic sublime, a particular visual practice that makes real "the tensions that arise from recognizing the toxicity of a place, object or situation, while simultaneously appreciating its mystery, magnificence and ability to inspire awe" (375). Through the perceptual disruption caused by the compositional incongruity, these images "alter the sublime response in order to encourage contemplation of the viewers' position within a polluted world" (375). Peeples cites the work of Edward Burtynsky as a representative example, and in many respects her analysis of these visual practices reveals the heterotopic quality of Burtynsky's images and the ambivalence experienced by the modern subject.

> The toxic sublime produces dissonance by simultaneously showing beauty and ugliness, the magnitude of the projects and the insignificance of humans, illustrating what is known of production and unknown of effect, questions the role of the individual in the toxic landscape while simultaneously eliciting the feelings of security and risk, power and powerlessness. (377)

Cultures have historically connected waste to dirt and impurity, and attempts to redefine and reassociate waste with the beautiful clearly prompt a state of dissonance. According to Peeples, the beautiful she attributes to Burtynsky's compositions disturbs the complacency of a public long accustomed to the practices of absence.

> In contrast, the unadorned and material substance of the altered landscapes may repel viewers. For those of privilege, waste and destruction are hidden from sight, whether under the sink, on the edge of town or shipped out of the country. Residuals from consumptive lifestyles are often intended to remain unseen from the consumer. The taboo against revealing the purposefully hidden can require an elixir, something to aid in the digestion of the ugliness of the subject matter. Beauty functions as that aid. (381–382)

The value of Burtynsky's images rests not in their capacity to aesthetically transform dirty matter into objects of beauty. Such alterity would simply reify the perceptions of absence that have fostered the ill-formed relations to waste. Instead, the images generate their rhetorical force by a perceptual disruption and the resulting search to understand how waste can be associated with beauty and how beauty can be aligned with the presence of dirt.

Other art forms, particular those that have come to be associated with Land Art and its various manifestations, have similarly drawn upon the heterotopic and unusual juxtapositions of objects to constitute spaces of modern dissonance (Thornes 2008, 393). Smithson's *Spiral Jetty*, which extends out into the Great Salt Lake in Utah, is most often characterized as the representative example of Land Art. While this example of Smithson's work has no direct connection to the presence or absence of waste, its relevance to my discussion rests in the way that the work advances the relation between the human and the other-than-human and the artwork's ability to expose the agency of the other-than-human. As the level of the Great Salt Lake changes, so too does the appearance of the *Spiral Jetty* jutting out into the water. Erosion from various natural processes likewise alters the shape and formation of the rocks and other objects that constitute the jetty. Thornes (2008) references Smithson's writings about *Spiral Jetty*, making note especially of the way the art object and its effects emerge through the configuration of human and other-than-human agency. "This powerful description of the evolution of *Spiral Jetty* reminds us that the work is a potent combination of the natural physical environment and the imagination of the artist and ultimately the spectator. The environment controls the emersion and the emergence of the piece, and it is forever changing in form and color" (401). *Spiral Jetty* materially asserts an uncertainty and unpredicted alterity, brought about by actions beyond human intervention. Stability and control brought about by the intentions and actions of a human agent are no longer the primary goals of the work of art nor are they even achievable; rather, revealing an uncertain world and the aesthetic that emerges through actions beyond the bounds of human intentions becomes the primary experience of the spatial configuration. On first thought, associating a toxic disposal area as a work of art and as an aesthetic experience may seem incongruous, but the efforts to constitute restored spaces as sites of natural beauty and efforts to hide landfills behind lines of trees speak to the prevailing aesthetic by which modern society has consistently attempted to beautify the dirtied and soiled. And the various awards given to restoration efforts that fall in line with the familiar landscape aesthetic of the beautiful further support the perceptual need to admonish the ugly and the dirty. But as my ensuing discussion will show, heterotopic configurations which juxtapose ugliness and uncertainty with the promises of a modern orderliness play a vital role in promoting public engagement with contemporary dirtied spaces and a resulting public performance of ecological care for these spaces.

THE AMD & ART PROJECT

Like many areas of the Appalachian United States, western Pennsylvania once provided a valuable resource, in the form of extracted coal, to promote the nation's economic and industrial development. Mining the coal also brought significant environmental harm to the region, however. Even as the coal reserves have been depleted and the mining operations have ceased, the harm to the environment continues, specifically in the form of acid mine drainage (AMD). "Seeping or surging from abandoned coal mines, AMD is the metals-laden water, often acidic, that coats stream beds with orange sediment, killing the bottom of the food chain" (AMD & ART 2016a, para. 1). To address the AMD that was occurring in Vintondale, Pennsylvania, and which was disturbing the area watershed of Banklick Creek, a creative partnership was formed that led to an innovative configuration referred to as the AMD & ART Project.

> We worked with many disciplines to destabilize the typical negative expectations of this region with a large-scale, artful public place that directly addresses the problems of AMD and much more. Beginning our work in Vintondale, Pennsylvania, in 1994, we've established a model of holistic renewal that brings the perspective of history to mix with the discipline of science, the delight of innovative design, and the energy of community engagement. (para. 2)

To address the contaminated water, a series of treatment ponds were constructed that consisted of limestone and various plant matter, with each pond playing a specific role in removing the metals from the tainted seepage. These treatment ponds and the processes by which the waste water is remediated became objects within a cultural heritage and recreational space. Alongside the treatment ponds were a multipurpose park and a heritage trail with various historical markers documenting the region's industrial past. In fact, the wetlands, where the water is eventually dispersed after its movement through the various treatment ponds, illustrated the innovative connections between the various disciplines and practices that planners saw as constituting this heterotopic space.

The wetlands were designed as more than just representative examples of the human potential to remediate disturbed landscapes and restore them to notions of the Edenic and the beautiful. The wetlands also were planned to convey the history of the region and have been referred to as "History Wetlands:"

> Here, where black bony, or waste coal, once barely supported scrubby grasses and stunted trees, the new wetland environment is attracting a variety of birds and wildlife, as well as human visitors. It is also within in [sic] these wetlands that the past of the site is most tangible, as the footprints of the old colliery buildings rise from the wetlands as ghostly reminders in the landscape. Our hope was to bring Vintondale's history back to the surface, to celebrate both its proud past and its future commitment to environmental improvement at the same time. In addition, all three of the site-specific installations—the Mine Portal, the Great Map, and the Clean Slate—are within the History Wetlands. (AMD & ART 2016b, para. 2)

The configuration materially juxtaposes the stones from the abandoned industry, the waste in the form of the AMD, and the pastoral, offering in this heterotopic space a clear representation of the dissonance that marks contemporary reality. As the planners of the AMD & ART space proposed, waste, in the form of acid mine drainage and the material footprints of the lost industry, was not matter kept from sight nor invisible to visitors. The detritus has not been transported away or directed to a publicly inaccessible location. Instead, waste became a highly present participant in the making of the incongruous space and the resulting novel perceptual experience of the visitor.

Due to its innovative integration of waste, its novel approach to cultural heritage, and its intended partnership of science and art, the AMD & ART project drew a significant amount of critical acclaim. In his interview with T. Allen Comp, the manager of the project, Erik Reece moves the conversation about the project beyond whether the configuration is a work of art and instead directs attention to the value of the space in relation to promoting environmental consciousness and care: "Science can change the water chemistry, but for Comp, it is art and history, combined with the science, that will ultimately change people's minds—will change the way we think about an industrial economy that is destroying the very ecosystems that sustain us, and all life" (Reese 2007, para. 32). The transformative potential Reese attributes to the AMD & ART space echoes the effects Peeples (2011a) ascribes to the toxic sublime found in the images of Burtynsky.

> [A] comparison of the self to the toxic sublime raises questions of complicity, producing an internal reckoning (at least initially) as one measures one's life choices against the sites of destruction. The recognition of a connection to these toxic places is the important step to understanding the need for alternative resource and waste protocols and decision-making. (388)

Yet in 2021, the AMD & ART project is far from the configuration intended by its creators and supporters, and its capacity to invoke the change in public consciousness hoped for by Reese and others has remained unfulfilled. What were once holding ponds designed to treat waste and to offer a novel approach to encountering and understanding the dirty matter are now empty pits, overgrown with weeds and other plant life. In an unpredictable and unintended example of Land Art, the AMD & ART project has become transformed, reshaped by the incursions of unexpected actors and, perhaps most important, by the waning responsibility and capacity of humans to manage and care for the space. While its initial novel configuration offered the potential to revise perceptions of space and waste, the reshaped AMD & ART space now offers compelling testimony of the unmistakable importance and role of human participation within a dirtied world and the impacts of an absent public, human actor.

In scientific terms, the AMD & ART project has been classified as a high risk failure, since "the design was poor and did not meet design standards known at the time" (Rose 2013, 4). As a result, the passive water treatment systems designed to treat the acid mine drainage have failed, resulting in effluent that retains acidity. But the configuration has also failed due to the alteration of its original arrangement, changes attributable to the lack of maintenance and management of the site. Rose's analysis of the AMD & ART project reveals that no organized oversight of the configuration exists with public efforts to engage with the performance of the space currently occurring strictly by happenstance.

> When visited on 7/8/13, two teenagers approached us. They say that the inflow from the mine into the pipeline was cut off for several years by plugging of the inflow grating and a break in the pipe, which is on the surface part of the way. In June 2013 they cleaned the inflow grating and put a "bandage" on the broken pipe, resulting in renewed flow into the system. Their actions were not part of any organized group—they were completely out of their own interest. The original agreements are said to involve the town of Vintondale for maintenance, but this is apparently lacking. The system badly needs a local or other group to monitor and maintain it, and to renovate it to be large enough to handle the load. (Rose, Appendix B, 3)

The promises expressed by Reese (hopes that have become quite familiar to environmentalists) have failed to materialize from the configuration, with the only instance of environmental care and responsibility emerging through the unstructured actions of teenagers. The blame for the space's failure to transform publics into more environmentally conscious consumers should not be strictly aligned with the botched application of science or the challenges to perceiving the innovative design, however. More significant limitations to constructing the desired human subject exist. If a key component of the juxtaposition included the "energy of community engagement," the resulting alterity of the site clearly demonstrates the lack of this central actor's participation. Viewing the site may have offered, as proposed by Reese, an opportunity for publics to become moved, primarily through the act of passive spectatorship. The proposed participation and energy of community engagement offered another means to move publics, one that entailed being more present with waste and the entanglements that constitute contemporary society's efforts to dwell with the dirty matter. The failure to collectively act and to collectively engage with and care for dirtied space can be traced to the familiar practices of making waste absent, particularly those modern, specialized human actors and institutions that have consistently made waste absent for publics. The actions by the teenagers to be present with the waste and to become entangled within and responsible for this contemporary configuration of dirty matter indicate the potential for such novel configurations and spaces to become publicly provocative. Therefore, the potential of heterotopic spaces such as the AMD & ART site to rearrange public perceptions rests not in the way they may provoke affects of guilt and remorse so as to change habits of consumption; rather, their promise rests in the way they may provoke a new responsibility toward the continuing maintenance and care for the present, contaminated configuration of the planet. Such responsibility must move beyond relying on specialists and those others typically assigned to waste management since this distribution of responsibilities only serves to further promote waste's absence and public inaction and inattention. If caring for the commonly referenced natural world served as a major component of modern environmentalism, then caring for the contemporary dirtied world and coming to understand its often incongruous arrangements will need to be a part of posthuman environmentalism.

THE SUBLIME WELDON SPRING SITE

After crossing the Missouri River driving west on Interstate 64, one starts to reach the outer northwestern ring of surburban St. Louis. On the southern side of the interstate sits the Weldon Spring Conservation Area, roughly 8,400 acres of mixed forest, pasture, and wetlands. With numerous hiking trails, fishing ponds, lakes, and access to the Missouri River, the Conservation Area provides residents of the populated St. Louis metropolitan area an opportunity for many familiar forms of outdoor recreation, including the opportunity to hike to the highest location in St. Charles County. But the perch from this vantage point comes not from a natural, geographic formation but rather a seventy-five-foot-high concrete containment cell housing 1.13 million cubic meters of radioactive and toxic waste, all remnants from a military weapons production facility that once stood on the location. This containment cell is a part of the Weldon Spring CERCLA Site, a remediated Superfund location. The contamination of Weldon Spring can be traced back to the 1940s and the government's acquisition of what was at the time private and incorporated land that was converted to the Weldon Spring Ordinance Works. During the 1940s the Ordinance Works produced DNT and TNT for use during WW II. During the 1950s, the site was transformed into the Weldon Spring Uranium Feed Materials Plant, where the Atomic Energy Commission produced uranium metals for the country's nuclear arsenal. The residual toxic materials from the uranium processing and TNT production were stored on-site in a large quarry and various pits. Given the lack of oversight during the processing and production, the soil, groundwater, and many buildings used to manufacture the material became contaminated as well. In 1985, the Department of Energy took control of the property and began cleanup and remediation activities, with the contaminated soil, building materials, and toxic matter now housed beneath the containment cell.

In addition to hosting the highest place in St. Charles County, the Weldon Spring Site is also home to an interpretive center, which contains numerous artifacts and exhibits that document the transformation of the location and the contributions of the weapons production at Weldon Spring to the nation's defense during World War II and the Cold War. The site is also home to the restored landscape of Howell Prairie, an example of what some may label as conventional remediation practices since it offers visitors an encounter with some of the flora native to this area of Missouri. In addition to the cultural heritage and biological experience, Weldon Spring also offers recreational activities by way of the Hamburg Trail, which meanders through the center of the site. The Hamburg Trail also connects to a number of other local paths such as the Bush Trail and the Katy Trail that run along the banks of the Missouri River, further extending the opportunity for various leisure activities. The juxtaposition of these various places generates an initial sense of displacement given that practices of leisure, tourism, and recreation are not typically juxtaposed with toxic waste facilities. In many ways, such disharmony and incongruity rests at the center of the OLM intention to configure such a space, given that one of the primary stated aims of the site is to foster public trust in the process of remediation by way of offering such transparency. In this regard, the site attempts to address one of the central problems of risk society, namely the ability of the public to trust those institutions that draw from techno-scientific processes to make waste absent. As a result, one interpretation of the meaningfulness of Weldon Spring emerges through its capacity to generate trust by offering such a public presence of radioactive waste.

As Kasperson and Kasperson (2005) note, trust, especially in situations of risk, is dependent on the perception of bias, given that a public expects that those overseeing public safety "must be able to conduct its activities uncompromised by any hidden agenda or undue influence by particular interests" (23). Trust is also a product of technical competence, the belief that the "agency has the requisite expertise and information to carry out its mission and to protect human health and safety" (23). While much work has been devoted to understanding trust in relation to risk communication, these investigations have largely been informed by modern ideology, associating trust with the performance, both technical and moral, of human actors and institutions. Through this modern lens, specialized institutions are assigned the task of ordering the placement of waste and securing the public from any harm, with trust in these institutions being important to establishing the needed security. Gross (1994), however, in his work on the rhetoric of science, moves us toward recognizing the larger assemblages in which society establishes moral orders, a process he labels as a rhetoric of accommodation. Modernity moves us to associate responsibility to the actions of humans. In the case of drunk driving, for instance, "it is the cultural consensus in the United States that the individual driver is the primary agent" (14). But Gross offers an alternative way to perceive agency and accountability. "A reasonable approach to automobile safety, then, might involve distribution of moral and legal responsibility among the manufacturer of automobiles, the builders of roads, the manufacturers and purveyors of alcohol, governmental authorities, and the individual driver" (14). Gross thus prompts us to consider trust and accountability as attributable to more than just the actions of an isolated, human subject, outlining how responsibility may be diffused throughout heterogeneous objects and practices that constitute a configuration. And while not explicitly referring to new materialism and waste, Gross offers a preliminary path by which to understand the larger network of actors and relations involved in generating trust and securing public order in situations where waste and toxins may present risk.

From a distributed perspective, the capacity to manage risks from contemporary toxins becomes a product not just of those specialized human actors and institutions but also the performance of the other entangled and distributed human and nonhuman objects, actors, and practices. Understanding posthuman risk, therefore, requires perceiving connections among the dispersed yet related material bits and pieces, an essential step that leads to revealing the extent to which all this matter holds together and alleviates any sense of danger. From this perspective,

> risk is neither a property of the human or nonhuman world but arises from the interactions between them and is performed by the complex ensembles they constitute. A condition of risk exists when the performance of an ensemble varies or deviates from that intended so as to result in unwanted, deleterious consequences. A key concern here is to provide an account of these dynamics and to explain how conditions of risk arise and might be ameliorated. (Healy 2004, 284–85)

Subsequently, the risk a toxin poses does not result from an inherent danger within the isolated matter but rather through the particular arrangement of objects in which the dirty matter is emplaced. Such a view, however, does not neglect the capacity of waste, a potential that is particularly manifest in the ability of dirt to move and unexpectedly remake spaces and relations.

Waste may have the capacity to move, but its potential to disrupt and to generate risk emerges through the novel relations and spaces that are created through its unplanned and unrestrained trajectories. Therefore, the capacity to limit waste's movement and to disallow it to move among and into new relations informs the contemporary practices of risk management.

Land Art may reveal the tenuousness of human intentions and actions to preserve and maintain some material and spatial stability, but spaces devoted to emplacing and encountering waste must operate through some measure of, as the abandonment of the AMD & ART site shows, human action and responsibility. A new materialist orientation to waste seems to place humans in a precarious position in which human agency may seem incapable of providing safety from the risks of waste and toxins. An acceptance of the agency of the other-than-human, however, need not equate to resigning oneself to happenstance or an acceptance that humans are incapable of shaping the world for the better. The agency of the other-than-human and the continuing threat of unpredictability and novelty do not equate to a determined, incapacitated human subject; rather, this uncertainty directs attention to the creativity and increased role of the human agent. This perception of risk reshapes human participation within a dirtied world to include not just acts engaged with limiting the production of dirty matter but also to acts involved in caring for an altered landscape that extend beyond the conventional practices of remediation and restoration.

To explore this potential for public participation, I will draw from my encounter with Weldon Spring and suggest how the experience evokes the sublime, a perceptual state that emerges through the novel and incongruous juxtapositions that characterize posthuman heterotopia. The sublime experience occurs sequentially, as Oravec (1982) describes, as the initial encounter with the sublime object "produces sensations of overwhelming magnitude and quality" (219) which the subject is not immediately able to comprehend. That initial cognitive dissonance leads to the second affect, a state in which "the sense of self is diminished and the individual feels a separation from, or lack of control of, the natural environment" (219). But the sublime experience does not end with the human subject acquiescing to human inferiority or to perceptual incapacity. The subject moves "past the feeling of inferiority to a sensation of exaltation, or at least wonder, at the relative grandeur of the object compared to the self" (219) and, in the case of the technological sublime, to feelings of human superiority and the prospects of a better world. An encounter with the toxic sublime as in the images of Burtynsky, however, moves the subject on a different, final trajectory, away from the utopian possibilities of human innovation and toward "alarm for the immensity of destruction one witnesses," a subjectivity that affords one "to question the personal, social and environmental ethics that allows these places of contamination to exist" (Peeples 2011a,

380). The affective state of distress and anger engenders a critical relation with the toxic object.

> Envisioning one's self as separate from the contaminated places, unaffected or untouched by them, a practice encouraged by waste disposal policies and systems, provides no imperative to act. When one measures the self against these sites, it is not necessarily an evaluation of moral character or spiritual strength, as would be the case with nature. It instead requires a confrontation with our consumptive habits, what we buy, where we buy it, what organizations and industries we directly or indirectly support, and how those choices are influential in creating the sites we see. (386–387)

In this familiar encounter to contamination and despoiled lands, reconciliation of the otherness prompted by the toxic sublime occurs through the invocation of guilt, tied especially to one's complicity in disturbing the purity of the natural. The distress of the ruined landscape and the resulting shame brought about by one's complicity in generating such disturbances produce not just contemplative subjects but also engaged citizens who become transformed and revise their consumerist behaviors to prevent further environmental destruction. In this material-discursive order, the disturbed landscape, however, serves as little more than a hopeful prompt for those negative affects that may lead to the desired, ecological subjects. In other words, we accept the condition of a disturbed landscape as a means to prevent others from being similarly transformed.

The sublime experience that occurs through encountering the materiality of waste, as in the space of Weldon Spring, affords the potential to move subjects to ecological awareness and action, but the space works in entirely different and novel ways. The subjects' experience at Weldon follows the familiar sequence of the sublime, yet it is the persistence of incongruity that occurs as subjects encounter both the promise and uncertainty of human actions that generates the resulting engagement with waste. The encounter with waste matter does not invoke the subject to consider the "personal, social and environmental ethics that allows these places of contamination to exist," so as to then prompt one to revise practices of consumerism. Instead, an encounter with waste at Weldon Spring provokes the subject to consider the "personal, social and environmental ethics that allows these places of contamination to exist," so as to then prompt one's responsibilities in relation to the care and maintenance of these dirtied and incongruous spaces. Encountering waste at Weldon Spring generates a subject who must confront the posthuman reality of existing among toxic and active, other-than-human objects and who must also confront one's responsibility to understand and respond to the uncertainty that such spaces generate.

PUBLIC PARTICIPATION IN DIRTY SPACES OF PLAY

As I mentioned previously, a rationale for constructing Weldon Spring as a tourist destination rests on the assumption that public access to radioactive waste would result in an enhanced sense of transparency and subsequently lead a public to trust the processes used to secure the waste and maintain the needed level of public safety. If trust is a major constraint to alleviating a public's sense of risk, then providing a public with such open access to the site, according to the OLM officials, would result in a more knowledgeable and subsequently more trusting public body. But generating such confidence entails more than a public passively accepting the practices of specialized institutions. Access to the site provides the opportunity for a public to encounter the novel ecology of such a contemporary space and to understand the relations and performances of the various actors, including the value of a public's own actions with respect to the space's performance. Monitoring the site has been designated as a key activity by OLM officials, and the public has been granted some responsibility with this process. "One of the goals of the surveillance and maintenance program is to promote and facilitate public involvement" (OLM 2005, 2–3). Such public involvement begins with an understanding of the space's ecology, particularly in terms of identifying the various actors and relations that constitute the space, an understanding made possible through the discourse available in the Interpretive Center.

The Interpretative Center offers a range of material and written artifacts pertaining to the history of the site, yet a significant amount of this discourse is focused on the disposal cell, the pile of radioactive waste that is publicly accessible and that stands adjacent to the Center. While explanations of the cell's construction are provided, the available discourse largely centers on the performance of the enclosure containing the radioactive waste, stating that "the disposal cell constructed at the Weldon Spring site has been designed to deter the migration of contaminants and to remain stable for 1,000 years" (OLM 2015a, para. 2). Further efforts are made to promote the cell's stability, with descriptions of how the waste pile consists of "exposed surfaces engineered to resist long-term erosion potential and a precipitation event greater than has occurred in the recorded history of the region" (para. 2). These statements draw attention to the potency of the cell's configuration, an agency that is conveyed to site visitors: "The cover systems armor the top of the cell protecting it from erosion, infiltration, bio intrusion, etc. It consists of multiple layers including (from bottom to top) an infiltration/radon barrier of clay, a geosynthetic liner, a gravel drain, sand filter, and a mixture of cobbles" (OLM 2015b, para. 5).

The cover system, which might be perceived as an assemblage of seemingly abiotic and inert material, is, in fact, a configuration of active objects collectively performing significant acts. In the rhetoric offered by the Interpretive Center, the protection of the public is assigned to the arrangement of material, with the performance of these nonhuman agents portrayed in military language. The materials that constitute the cover provide armor and a shield to prevent infiltrators and intruders. Foes come not in the form of human enemies but rather in a range of nonhuman actors such as earthquakes, rain, wind, and even biotic, plant material. Through the discourse, publics became aware of the arrangement of the space, of the particular ecology and relations among the objects that allow for the waste to be contained and for risks to be minimized. Yet it is not until one voyages outside the Interpretive Center that an understanding of the ecology shifts to the sublime encounter and the resulting prospects for public interaction with the matter.

The incredulity that marks the publics' encounter with Weldon Spring gains its distinct affective character as visitors step outside of the Interpretive Center and onto the walking path that leads directly to the base of the disposal cell. From this position, the size and shape of the cell, which stands seventy-five feet high and extends across forty-five acres, dominates the visitors' gaze and field of view. Encountering the size and accessibility of the landfill further invokes the sublime, casting the human subject within a state of ambivalence. For example, my acquired knowledge of how to exist with risk asserts that the most effective way to alleviate the threat of toxins is through distance and absence. The discourse of the Interpretive Center only furthers this belief: "There are three concepts in basic radiation protection. They are: Time, Distance, & Shielding" (OLM 2015c, para. 13). Yet, during my visit, I, dwarfed by the size of the cell, willingly stand at the base of a pile of radioactive waste because of a curiosity invoked by the novel, dissonant juxtaposition but with some confidence in the objects and practices that may shield me from any harm. My encounter with the materiality of the disposal cell and the symbolic representations of its construction and performance do not move me toward a blind acceptance and appreciation of human superiority, however. While I acknowledge the capacity of the configured objects that constitute the disposal cell to provide some sense of safety, I also do not discount the vibrancy of the waste to push back against these assembled objects. And the discursive promises of my safety fail to further resolve my affective and cognitive incongruities, at least in terms of the intentions suggested in the site's official rhetoric.

> Walking to the top of the disposal cell and standing on the platform at the peak you will receive less exposure to radiation than you would receive

standing in your own backyard. The cover of the disposal cell consists of multiple layers, including the clay radon barrier, geosynthetic liners, sand, and crushed limestone rock surface layers. By its nature such crushed rock emits lower background radiation levels than soil. The combination of reduced radon emissions and lower background emissions from the crushed limestone rock result in lower overall radiation emissions on the cell surface than the average levels in clean soil in the St. Louis Area. (para. 25)

According to the site's rhetoric, standing atop the pile of contained radioactive waste presents less risk than an everyday experience of standing in my residential backyard. I am a bit skeptical, however, about this discursive effort to reconcile the perceptual incongruity generated by my bodily emplacement within Weldon Spring since the dissonance invoked by Weldon Spring mirrors the dissonance I find myself encountering in various quotidian, posthuman configurations. My backyard, while maybe emitting some radiation, does not contain a radioactive waste pile. But as I deliberate further upon the proposed attempt to reconcile the incongruity of my position, I must concede that I do not fully know what may be circulating in my backyard. If I am to honestly engage in the reality of a posthuman, toxic world, I must acknowledge that purity and cleanliness are no longer qualities of the spaces I inhabit, including my backyard. Hazards, beyond any radioactivity, clearly find their way into my residence, even beyond intentions. My yard is, in fact, a toxic stew that contains the excess fertilizer that flows into my backyard, a trajectory resulting from the rolling landscape that characterizes my residential street and my neighbors' efforts to eliminate pests and pesky weeds and to grow the greenest grass on the street.

My yard is also in the path of the oily residue that drops from parked cars onto my residential street and, during times of rain, moves along the curbside, with some of the gray, blue, and red tinted water sliding down my sloped driveway. I am unaware of what other invisible toxins and matter the rain may carry from the street and into my yard, ignorant not just of the dangers of these individual toxins but also incapable of acting in meaningful ways to prevent the toxic space that emerges from the collection of all the dirty matter that has found its way into my yard. I can monitor and surveil the movement of this waste as it flows into my yard, but I lack the means to participate in meaningful ways so that the waste does not flow at all. I may become part of a neighborhood effort to stop the use of pesticides and fertilizers; but while such collective action may stop the flow of these chemicals into my yard, the action would not address the chemicals that have already become a part of my residential space. The comparison used in the Weldon Spring Site's discourse, therefore, generates an effect, but not necessarily one that reconciles the incongruity and perceptual challenge that the configuration evokes. The

Interpretive Center's rhetoric may not make me feel at less risk from the nuclear waste I stand atop; instead, Weldon Spring reorients me to the posthuman world of risk and to the need to locate the perceptual apparatus and some course of action that may move me toward some means to better reconcile the fact that I am unable to completely distance myself from toxic matter no matter where I stand.

This movement toward reorientation is fostered by way of the site's alternative, material arrangement. Unlike the majority of landfills and remediated sites which have been converted to green, park-like space, the cover of the Weldon Spring disposal cell consists of a collection of symmetrically shaped and sized rocks, sloped at roughly a 45-degree angle. All the rocks share the same tone of gray, with speckles of occasional black on some individual rocks serving as the only point of differentiation. No form of plant or animal life is visible. Symbolically, the cell conjures images of a dead earth, the lifeless and poisoned landscape common to the apocalyptic narratives that prophesize a world tainted by nuclear contamination. The aesthetic of the disposal cell is more than symbolic, however, since the arrangement directly enacts the performance of the configuration. While the lack of any topsoil or biota may promote representations of death and destruction, the seeming lifelessness of the cell is a misinterpretation given that the configured matter is essential to preventing erosion and subsequent seepage of any contaminants. The landscape at Weldon Spring I encounter, therefore, does not promote death and lifelessness but rather a relational agency and contemporary ecological activity that works to ensure my safety. What I encounter as I face the disposal cell is not a familiar space of restoration "used to obscure or justify environmentally damaging practices" (Eden, Tunstall, and Tapsell 1999, 152). Instead, I confront a revealing, posthuman configuration, an assemblage that conveys to me the transformations and changes to the world and the novel posthuman ecologies that must become juxtaposed to ensure my safety. The cell at Weldon Spring introduces to me a new ecological, performative aesthetic.

But the configuration that constitutes the disposal cell is not entirely absent of biota, since Howell Prairie sits immediately adjacent to the pile of rocks. An initial interpretation of the OLM's rhetoric describing the prairie reveals echoes of the prevailing discourse of remediation that promise a once contaminated land's return to its pristine, Edenic past.

> An area known as Howell's Prairie was part of the historical range of prairies in St. Charles County and resided in the same location as the present day Weldon Spring Site. After site remediation, the Department of Energy restored the site to a native prairie habitat, which was deemed the ideal solution. This offered long-term protection around the disposal cell due to

the deep root systems that hold soil in place and resist erosion. The 150-acre prairie at the Weldon Spring Site was named Howell Prairie in keeping with the history of the land. (OLM 2015d, para. 4)

But the discourse strays from the conventional characteristics assigned to restored spaces, moving, as with the rhetoric describing the pile of rock of the disposal cell, to the performance of the prairie's particular arrangement. The lifeless configuration that constitutes the disposal cell maintains its shape and function, in part, because of the material objects and relations that constitute Howell Prairie. The prairie thus illustrates more than the human capacity to restore the landscape to some previous utopia, of a space of pristine and natural balance. The configuration demonstrates the human capacity to arrange a novel ecology whose performance includes helping secure the safety of humanity and other biota. The configuration of abiotic and biotic material that maintains order and public safety at Weldon Spring, a juxtaposition that revises perceptions of the apocalyptic and the utopian, thus raises prospects to reorient publics with respect to our aesthetic, performative, and critical relation to a contaminated world.

Reconciling that initial sense of incongruity fostered by the strange juxtaposition of the human body and radioactive waste, therefore, can only take shape through an ecological reorientation. And understanding the arrangement of the various objects within the space as indicative of an ecological performance as opposed to a representation of familiar aesthetics and environmental rhetorics affords the public the capacity to move beyond a passive, conceptual understanding to a participatory member within the assemblage. Publics, through their newfound ecological sense, become valued stakeholders and agents actively engaged in the performance of the configuration. As the OLM officials explain in the description of the monitoring plan for the site: "Stakeholders may comment on LTS&M Plan changes, provide review of DOE activities by reviewing documents and attending public meetings, informally monitor the Site and the site surveillance and maintenance program, and report concerns to DOE or regulators" (OLM 2005, 2–3). While engineers and other professionals with the requisite techno-scientific expertise will play central roles in inspecting and monitoring the Weldon Spring configuration, publics who visit the site and who gain the posthuman ecological understanding of the site's performance also have the capacity to contribute to these efforts. As the OLM advocates: "One of the goals of the surveillance and maintenance program is to promote and facilitate public involvement. Active public involvement helps DOE address citizens' concerns as well as provides additional surveillance input to DOE" (2–3). While the OLM coordinates conventional means of public participation such as community meetings and dialogue within its efforts to engage the public, the agency also directs the public to participate with the management of the radioactive waste in other ways.

> Contact information for the DOE staff responsible for implementing the Weldon Spring Site surveillance and maintenance program will be posted at the Interpretive Center and made available via the DOE-LM website. These communications will encourage the public to actively participate with DOE in the surveillance and maintenance process by reporting sightings or concerns such as visible changes to the cell cover, erosion, suspicious land use, damaged monitor wells, or vandalism. (2–5)

Through their active participation in surveilling and monitoring the disposal cell and adjacent grounds, the public has the opportunity to engage in performances that may reconcile the dissonance inherent to postmodern risk, particularly the uncertainty generated by the unpredictable movement and rearrangements of vibrant waste matter. Arranging waste within configurations of recreation and public access, therefore, may revise those perceptions and practices of absence that have consistently informed the populace's relation to dirty matter. The potential for a more engaged public arises not from the guilt and shame of admitting to one's complicity in producing the waste and the resulting tarnished and dirtied landscape. Instead, the configuration at Weldon Spring and even the intended space of the AMD & ART project work by affording publics the chance to acknowledge their responsibility for sharing in the performances that constitute being present in a dirty world. Making waste present affords the potential to make public acts of ecological care present. The realization of this responsibility becomes possible as society reorients itself to the novel ecology and arrangements that emerge in a dirtied world and develops the perceptual apparatus to understand the potential and limits of human capacity within such entanglements. The configurations that make waste present move us past the admission of guilt and direct us to identify how to now dwell with our dirtied selves and spaces. A public encounter of the sort afforded by Weldon Spring makes present the need to equip publics with a hermeneutical stance that readily admits to the incongruities and uncertainties of dwelling with waste; it affords the opportunity for publics to revise perceptions of ecological aesthetics and beyond the utopian dreams of modern techno-centrism and the apocalyptic visions of environmentalists' guilt and despair.

But moving publics toward such awareness and participation must also entail the continued transparency of institutions such as OLM. Trust, in a dirtied world, consists of more than a reliance on techno-specialists. In these contemporary spaces of paradox and heterotopia in which the public encounters waste, trust emerges from a transparency that admits to human constraints and challenges which, in turn, sponsor the need for continual and collective human participation to understand and shape the prospects for dwelling in a dirtied world. Such public participation has long been the concern of environmental justice advocates; but with the ever-presence of dirty matter, the need for public participation and a collective construction of trust now extend beyond the marginalized. If the encounter with the dirtied ecology in Florida requires a perceptual reorientation to better accommodate the various publics' perceptions of and responsibility to such entanglements, then a similar act of perceptual reorientation must occur to better accommodate the publics' perception of and responsibility to those other spaces in which waste is present and seemingly kept within some measure of control. This revised interpretive stance neither glorifies nor admonishes the role of techno-science in a dirty world; it rather admits to the role of technology and science while conceding some of its authority to other matter and to other participants, including publics. Public participation generated by the presence of waste within sites of recreation extends, however, beyond monitoring the stability of a local site's arrangement and posthuman ecology. If a space such as Weldon Spring can be transformed into a tourist destination and be afforded the needed institutional and public oversight to allay the incompatibilities brought about by waste's presence, then publics should also acknowledge and direct action so that other sites where waste is also present be constituted in similar ways.

RISK, PUBLICS, AND RELATIONAL ETHICS

Roughly twenty miles to the northwest of the Weldon Spring Site sits The West Lake Landfill Superfund Site, a 200-acre area that includes the West Lake Landfill and the Bridgeton Landfill. The area was initially a limestone quarry, and after the limestone was depleted, the quarry was converted to a landfill for municipal waste. As the USEPA (2021a) documents, however: "Two areas became radiologically contaminated in 1973 when soils mixed with uranium ore processing residues were brought to the landfill and presumably used as daily cover in the landfilling operation. An adjacent property has also been impacted, presumably by erosional migration of radiologically contaminated material from the landfill" (para. 1). While the national memory concerning its atomic past can easily recall such places as Los Alamos and Hiroshima, locations such as St. Louis, Missouri, are hardly recognized as being entangled within such war efforts. During the Cold War, Weldon Spring was not the only location in the St. Louis area that produced nuclear materials for the nation's defense. The St. Louis–based company Mallinckrodt Chemical provided a significant amount of radioactive materials for the nation's arsenal, and the company's work extended beyond Weldon Spring. A 2015 documentary, "The Safe Side of the Fence," attempts to add St. Louis to the national Cold War narrative and establishes the connections between various locations in St. Louis instrumental in generating nuclear material as well as those that have become reconfigured by its disposal. Barker (2015) mentions these various locations scattered around the St. Louis area in his discussion of the documentary:

> They may know about the cleanups in Coldwater Creek, the north riverfront downtown and by Lambert-St. Louis International Airport. Others may remember the uranium processing facility at Weldon Spring or know someone who worked at Dow Chemical in the Metro East. But local videographer Tony West figured many people probably haven't thought about all of them and how they're all connected to the world's first nuclear weapons. (para. 2)

Unlike viewing the documentary, a visit to the Weldon Spring Site will not provide an explanation of the extent of nuclear production and radioactive waste in St. Louis. The tourist encounter offered by Weldon Spring situates visitors only within the bordered space managed by the OLM. In many respects, the material-discursive rhetoric encountered by a visitor mirrors that encountered during a tour of the Manatee Viewing Center. In both cases, the tourist experience is informed by views of space as a bordered geography and by arranged objects contained within the marked boundary. As with the Manatee Viewing Center, moving beyond the borders of Weldon Spring so as to understand the relational qualities of dwelling with waste affords new opportunities to extend public participation and ecological action in a dirtied world.

As I have frequently noted, the intention to make Weldon Spring publicly accessible was to establish trust in the institution managing the site and the practices used to contain the waste material. A crucial means to establish the confidence in the site's stability occurs through reorienting public understanding of the arrangement of various material objects, specifically in terms of conveying the manner in which the material objects emplaced within the arrangement actively constrain the movement of the waste. Armed with this posthuman, ecological understanding, I, as well as the numerous other visitors to the site, can embark upon informal monitoring of the arrangement. The knowledge of this new ecology and the opportunity to publicly encounter and even monitor these arrangements position myself and other visitors as active participants dwelling with dirty matter. Locating such an encounter with waste in middle-class suburbia and juxtaposing such an encounter with middle-class acts of leisure thus offers an opportunity to reconsider absence and responsibility in the context of environmental justice.

As the presence of the dirty water along Florida's coasts illustrates, it is becoming increasingly difficult for all economic classes to remove themselves from spaces of contamination. But even with the ever-presence of waste, I am not suggesting that the unequal distribution of the risks from toxins is no longer a matter in need of attention. Concerns for environmental justice will continue, yet the ubiquity of waste and the emerging practices enacted to dwell with the dirty matter provide a means to reorient our perception of and approach to ecological injustices. For instance, given the movement into and the entrenchment of waste within Florida's waters and the presence of waste in such public, remediated sites as Weldon Spring, attention will need to be diverted away from the capacity of institutions to move dirty matter as a means of establishing public absence. And movement as a means of disposal has always been cast as unjust, given the frequency of disposal sites within or proximate to marginalized and disempowered publics. But in the fouled, posthuman landscape, equity can no longer be solely measured by one's capacity to escape from dirty matter and dirtied spaces; in posthuman heterotopia, equity can also be measured by the availability of practices all citizens have to dwell in dirtied spaces.

For example, while Weldon Spring and the West Lake Landfill both constitute St. Louis's radioactive past, the encounter with waste while atop the disposal cell in Weldon Spring is far different from the experience of a resident living in close proximity to West Lake. While I encounter the sublime strangeness of the disposal cell and intellectually reflect upon the ambivalence the configuration generates in this tourist location, residents living in the presence of West Lake Landfill experience a disturbingly different state of uncertainty. For residents near the West Lake Landfill, their vulnerability is generated by the existence of a subsurface smoldering event (SSE).

> In December 2010, Bridgeton Landfill LLC notified Missouri Department of Natural Resources (MDNR) that it found elevated temperatures in its south quarry. The company and MDNR eventually determined this to be an underground smoldering event. A subsurface smoldering event (SSE) is a high-temperature, self-sustaining (without the need for oxygen), chemical reaction that is consuming the buried waste (accelerating decomposition). (USEPA 2018, para. 2)

Local residents thus face the prospect of radioactive contaminated leachate moving from the West Lake Landfill and, should the SSE expand beyond the Bridgeton Landfill, the possibility that "such heat generating reactions can increase the potential hazards and the likelihood of an incident" (Feezor 2020, 4). After considerable efforts by local advocacy groups such as JustMomsStL, the EPA has taken steps to extract some of the radioactive waste from the West Lake Landfill, yet the "smoldering event" and the uncertainty it creates persist.

Given the uncertainty, state officials monitor the movement of the SSE and any potential radioactive leachate, but such actions do not foreclose the potential of a significant disturbance to the local community. Therefore, plans have been formulated for a range of possibilities, conveyed in a publicly available incident report. The document

> describes plans to prevent incidents, required protocol for initial incident emergency calls, coordination of responses, and resumption of normal activities (in case of interruption). As used throughout this plan, the term "incident" means a situation that is non-routine or is anomalous and which poses a threat to the health and safety of on-site personnel or the public, or which may develop into such. (Feezor 2020, 2)

While such planning and preparation are important to the safety of local residents, the admission of the possibility that non-routine and anomalous events may develop illustrates the uncertainty that embodies this particular configuration, a dissonance not readily perceived by visitors recreating within the Weldon Spring site. The public acts of informal monitoring that can take place during tourist recreational activities at Weldon Spring are clearly dissimilar and ethically unequal to public attempts to monitor the West Lake landfill.

More than offering a historical account of St. Louis's relation to nuclear weapons, a configuration that includes Weldon Spring and West Lake thus brings to the fore novel, normative questions of equality with respect to a public's relation to waste and the responsibilities publics have to dwell with waste. The fact that the West Lake Landfill is not constructed as a tourist destination nor is configured with recreational objects and activities indicates the need to view environmental justice not just in terms of the unequal placement of waste but also the unequal prospects for dwelling with the dirty matter in a posthuman, sullied world. While extending the space of the Manatee Viewing Center beyond its geographic borders so as to include the range of actors who constitute its stability and instability may provoke new ways of perceiving the relationality that encompasses contemporary environmental issues, so too would reconfiguring the tourist space of Weldon Spring to include the West Lake Landfill. Such a reorientation to space enables the emergence of a posthuman ethic of care, a capacity that is "attentive to the needs of both self and other" (Baumlin 2020, 2). As with other posthuman performances, this ethic of care emerges relationally, through the meeting up of seemingly disparate actors, spaces, and practices and the resulting responsibility to the others we find ourselves newly emplaced among. While adopting the relational perspective to dirtied space may reveal the range of actors and practices who are complicit in the production of these dirtied sites, we can also draw upon the relational to identify the potential for novel practices and performances to enable a more ethical way of being present with waste within reconfigured and uncertain spaces. The relational view can position us within the familiar position of casting blame and even casting responsibility for the dirt upon a host of what may otherwise be hidden actors. But the relational view as proposed in this chapter moves beyond that familiar spatial figuration to propose that relationality requires publics to become perceptive as to how their capacity to dwell with waste differs from the capacity of others. A contemporary relationality thus becomes a matter of shared and equal prospects for living within dirty space and one's responsibility to such an endeavor.

5

✢

The Ethics of Agency in a Dirty World

Approximately 100 miles west of Chicago, nestled at the bottom of a bluff at what is known as the great bend (the Illinois River's ninety-degree southern turn), sits the village of Depue, Illinois. Distanced from the spreading development of the Chicago metroplex, the approximately 1800 residents of Depue find themselves remotely situated within the Illinois River's floodplain, with the town's namesake body of water, Lake Depue, lying between the village and the river. During the last weekend of July, however, the quiet isolation of the rural village is transformed as up to 30,000 people converge for the Lake Depue National Championship Boat Races. Race fans, current and past residents, and those living in nearby towns gather on the banks of Lake Depue to watch the races, eat a variety of festival and regional foods, visit the beer gardens, enjoy entertainment acts, run in a marathon, and partake in the festive remaking of the lakeshore. Yet since 1995, the festival goers have come to witness more than just championship boat racing; they also experience what the New York Times refers to as the "dirtiest boat race in America" (Florio and Shapiro, 2016). The characterization is not due to any poor sportsmanship exhibited by the racers but rather to the fact that the race and the entire festival grounds sit atop an unremediated Superfund site.

Like the many other rural towns that border this stretch of the Illinois River, Depue once had a steady industrial base that provided residents with consistent employment and a middle-class standard of living. In Depue, the primary sources of such social and economic welfare came from a fertilizer plant on the city's northern border and a zinc smelter located on the far eastern edge of the community and a few blocks north of

Lake Depue. But similar to the changes that occurred to other rural, Midwestern villages during the latter half of the twentieth century, Depue witnessed the loss of its industrial base when the fertilizer plant and zinc smelter shut production in the late 1980s. And as with other deindustrialized communities, Depue also found itself with a row of empty industrial buildings, visible remnants of its past economic base. While the vacant smelter production buildings have been razed, the village's industrial past is now remembered by way of the invisible toxic contamination, in the form of various carcinogenic metals, spread out among what the EPA designates as five operational units (OU). Lake Depue is designated as OU-5, with its concentration of metals including arsenic, cadmium, zinc, and lead resting within the sediment at the bottom of the lake.

Similar to the challenges faced by publics residing in other Superfund sites, the citizens of Depue have dealt with numerous constraints in their attempts to prompt the principle responsible parties (PRP) accountable for the contamination and the Illinois Environmental Protection Agency (IEPA), who is charged with the cleanup, to remediate the lake in a timely manner. The inability of residents in such toxic environments to provoke and shape action to address the contamination is well documented, a rhetorical constraint resulting largely from the reliance on specialized expertise to assess and remediate the risks as well as the political influence of the PRP. Scholars and advocates have worked to remedy such disproportionate participation through various means, and in Depue, residents have been able to partner with the Northwestern University Law Clinic and specialized scientific experts as a way to gain some agency. In these collaborations, rhetorical agency is aligned with human-constituted epistemological and institutional structures and the ability of a human actor to affect other human actors by way of one's proficiency with sanctioned ways of knowing and speaking. Yet recent studies related to the ontological and relational turn have attempted to recast notions of agency. Instead of attributing agency to just human actors, new materialists have advanced the possibility that nonhumans and material objects have the capacity to affect practices and other actors, primarily through the emergence of novel spatial configurations and assemblages that human and nonhuman actors find themselves emplaced within. Agency, from this new materialist perspective, is not an individual, human trait but rather a relational and material (both human and nonhuman) emerging capacity.

This chapter takes up the new materialist work on agency, with special attention directed at the assertions inherent in new materialism that novel configurations prompt the potential for a renewed commitment to environmental care. If agency, as advanced by a new materialist approach, is constituted by the configuration and relation of human and nonhuman objects, then a reconfiguration of these objects and the subsequent relations among them may offer residents entangled within spaces of environmental injustice the capacity to generate actions to better their environs. In other words, a novel rhetorical situation affords the means for new and more ethical rhetoric and ways of living to emerge. But as my discussion of the dirtied waters of Florida illustrate and which I will elaborate upon in this chapter's discussion of Lake Depue, configurations that emerge in a contaminated world may not directly invoke the transformation of an ecological human subject. These spaces, true to their heterotopic nature, disrupt the prevailing orders and rhetorics of environmentalism; and because of their inherent incongruity, these configurations force a reorientation of the interpretive frames by which human action and environmental care may become reconciled. The unpredictable, heterotopic configuration that emerges in Lake Depue may indeed offer publics some novel sense of agency, but in turn this same configuration invokes the need to reconsider the ways we deem such agency and the corresponding human actions as promoting environmental care.

ENVIRONMENTAL INJUSTICE AND AGENCY

As of October 1, 2021, the EPA had identified 1,322 sites on the Superfund list (USEPA, 2021b). A number of these contaminated areas can be characterized as sites of environmental injustice, since many Superfund sites are adjacent to or located within minority and low-income residential areas. Depue can be characterized as not just a Superfund site but also a location of environmental injustice given that the median family income of Depue residents falls below the average of Bureau County as well as the state of Illinois (NCICG 2014, 20), that just under 12 percent of those living in the village had obtained a college degree (NCICG, 2014, 9), and that the town is geographically remote and subsequently economically isolated. Due to their marginalized economic and educational status, residents living within Superfund sites such as Depue are marked by the inability to act upon and make changes to their world, actions that would initiate and direct steps toward any remediation and cleanup of the toxic contamination. While the primary responsible parties (PRP) assigned legal accountability for the toxins and state agencies have made some effort to remediate the contamination in Depue, including the cleanup and monitoring of the smelter grounds and the ongoing removal of contaminated soil in residential areas of the village, the pace of such actions has been a point of contention and illustrates the residents' limited agency. "Environmental officials say the pace of cleanup in DePue is typical of Superfund sites in Illinois and across the country, where it can take years, even decades, to get polluters to pay or find federal money for cleanups—

not to mention simply doing the work" (Webber 2011, para. 12). The slow pace and institutional malaise that characterize Superfund cleanup processes, including those of Depue, may be attributed to the program's lengthy bureaucratic procedures, which involve a range of ecological assessments undertaken to determine the toxicity of a site and a series of feasibility studies used to identify courses of remediation and the possible reuse of the previously contaminated land. But such institutional inaction can also be ascribed to the significant political and legal influence of the PRP and, as many have argued, to the practices that privilege technical and scientific knowledge to assess and address the health risks that contaminants and industrial toxins may pose.

Subsequently, those without the ability to generate claims based on such an epistemology are left with little to no agency to contest institutional findings or to generate accepted alternative actions that may more immediately promote a village's health and well-being. As Fischer (2005) notes, "regardless of their political strength, interest groups and social movements without access to expertise can scarcely participate in the policy process, certainly not effectively" (23). Beck ([1986] 2011) offers a similar characterization, contending that those affected by risk "are becoming *incompetent* in matters of their own affliction" (53). Any efforts to generate action and change by a public that run outside of the established scientific epistemology are often characterized in derogatory ways, furthering the ideology that the only viable subjectivity on the part of the public is to become educated about risk by way of the prevailing notions of scientific reasoning. Lacking the requisite engineering and scientific knowledge, the public, when voicing concerns about risks and the assessments of technical experts, is labeled as irrational and emotional and as exhibiting, "the indecorous voice" (Cox 2013, 255). Subsequently, any public performance and identity that advances alternative perceptions and knowledge but that falls outside of the sanctioned ways of knowing and speaking is delegitimized. From this perspective, a quest for public agency within environmental justice sites finds itself confronting a seemingly rigid and durable institutional structure constituted largely by technoscientific knowledge.

To address the public's inability to act and to generate a more authoritative public identity, researchers and advocates have directed significant effort to challenging the privileged epistemological and rhetorical practices of science. Beck ([1986] 2011), in his landmark work on risk, has laid the groundwork for such alternative forms of reasoning and rhetoric by arguing that definitions of risk based on science and Enlightenment forms of reasoning are insufficient since such reasoning has not been able to control and effectively address the risks modern industrial practices have wrought. The limitations of scientific reasoning have led to a later stage of modernity and to the need to discover an alternative means to conduct human activity, which Beck labels as reflexive modernization. Specifically, reflexive modernity can be characterized by the movement toward a more dialogic political process as opposed to one constrained by expertise. Essential to Beck's later modernity is the opening up of risk politics to the public and the legitimization of knowledge and practices not typically associated with scientific and modernist forms of rationality. "We need a new symbolic ordering of risk-perceptions and articulations as the nature of some of the contemporary hazards, which transforms social risk relations of definition at a structural level" (Beck 1998, 199). Lash (1994) has likewise questioned the merits of the scientific method to assess and remedy risks and sees agency as the key issue in relation to risk society and the latter stage of modernity. For Lash, "the reflexive modernization thesis has for its core assumption the *Freisetzung* or progressive freeing of agency from structure" (119). The increased use of alternative knowledges by the public and social movements to challenge scientific definitions and assessments of risk can be seen as evidence of these theoretical assertions.

Fischer (2005), for example, proposes the value of cultural knowledge, defined as "knowledge pertaining to a local context or setting, including empirical knowledge of specific characteristics, circumstances, events, and relationships, as well as the normative understandings of their meanings" (194). Whereas the scientific approach to risk assumes a level of objectivity and thus universality, cultural knowledge situates practices and knowledge about risk within specific local and cultural spaces.

> In important respects, these contextual demarcations offer a valuable check against scientific tendencies to emphasize generalized knowledge. But this local knowledge offers more than just the opportunity of filling gaps in a scientific paradigm. It also involves alternative ways to conceptually organize and understand nature. (207–8)

Pezzullo's (2007) exploration of toxic tours illustrates the way that contextual and local knowledge can be a means of public agency and a source to generate awareness of environmental hazards. These tours are led by residents who live among the risks of industrial facilities and offer an alternative account of the experience of living in such close proximity to toxins. The excursions within contaminated sites foster a material and emotional effect on those touring the places of contamination, as the tourists become present to the daily smells, sounds, and bodily sensations common to such industrial spaces and come to feel the anxiety, worry, and danger that the residents of these sites of environmental injustice experience as a part of their daily lives (9–10). Others have likewise argued for an alternative epistemology and rhetoric by advocating for the legitimacy of emotions such as anxiety and fear, a move that subsequently challenges the prevailing ideology that casts the negative subject positions upon the public. Summarizing these affective approaches, Zinn (2008b) contends that "fear and anxiety caused by the socioculturally mediated reality of dangers and the experiences of bad risk management can become a source for political power or even a new political subject" (196). And Tulloch (2008) references Lash's call to acknowledge the aesthetic as a form of alternative knowledge, specifically in relation to the "terrible sublime (which is the gut-wrenching feeling of fear and trembling when we face some horrendous experience in our own life, or, more often through art and popular culture)" (149). Regardless of the various approaches advanced, granting the public greater access to deliberations on risk and promoting an enhanced sense of agency have become central elements of contemporary relations to modern industrial toxins.

Informed by the social construction of knowledge, numerous scholars have explored the way that rhetoric can be a means to redefine notions of risk and the risk subject. Articulation, as it corresponds to linking discursive elements to generate novel meanings, has been an important resource guiding this research. As I noted earlier, DeLuca references the transformation of Lois Gibbs from housewife to effective activist as a result of her entanglement within various discourses and the capacity afforded by the entanglement to form new linkages, relations, and meanings. Similarly, in her analysis of the placement of a waste incinerator in Spokane, Washington, Peeples (2011b) describes how those opposing the siting were able to link terms to not only foster a particular identity but to also generate an understanding of and a potency to their concerns: "community members used the traditional notion of being 'downwind'—that used by hunters and sailors—to explain the location and contaminant aspects of the dispute. In this, the active citizens were enacting a simple articulation—the linking of 'Spokane' to being 'downwind'" (254). In this instance, downwind has become reconceptualized, associated with populations who find themselves in spatial proximity to the detrimental effects of various industrial practices. "The original use of 'downwind' was as a nautical or hunting term used to establish position around a point of reference. With its association with nuclear radiation, it reversed its connotative meaning from that which was neutral or even favorable, such as being downwind from a threatening animal, to one that was unequivocally negative" (249). The environmental justice movement can be seen as constituted through other acts of discursive articulation. Schlosberg (2013), for instance, proffers a definition of the global environmental justice movement based on linking the elements of equity, recognition, and participation. A central articulation that initiated the emergence of the environmental justice movement was its rearticulation of the "environment." Whereas the primary linkage of the environment was to those pristine spaces populated by nonhuman inhabitants, environmental justice advocates established novel associations with the environment, revising the term to also include places "where we live, work, and play" (Novotny 2000). And while the articulation that has established the environmental justice movement has been crucial in bringing to light and even addressing such concerns as the inequity of the placement and presence of toxic waste, the movement's linkage with and emphasis on those places of the human have resulted in the negation of the nonhuman and an environmental movement where the "wilderness is invisible" (DeLuca 2007, 30). Granting a place for the nonhuman in environmental justice entails more than just acknowledging its existence and adopting a modern sense of care for this other; instead, the presence of the nonhuman involves a reorientation to this other so that matter becomes a recognized active participant in reconstituting the spaces in which humans are emplaced and the practices and ethics humans enact.

While efforts to enhance public participation have focused on enabling the human to enhance its rhetorical proficiencies and to better understand the way that new linkages within discourse can create new worlds, an acknowledgment of matter's vibrancy leads us to attend to the novel linkages fostered by the movement of the nonhuman and the resulting (re)placement of humans within these reconfigurations. Linkages and new meanings may result not just by way of discursive elements but also by way of the novel arrangements of material objects and practices that result from matter's rearrangement of space. Subsequently, we should turn our attention toward the articulation of the material and the rearrangement of human and nonhuman objects to more fully understand the emergence of agency within sites of environmental injustice. This chapter, following Wells et al. (2018), thus sets out to explore the potential of attending to such creative and ethical intra-action:

> An ecological approach asks whether, instead of oscillating between social and environmental well-being, we might seek a more substantial recognition of the entwinement of the two. . . . And how might publics come to such recognition? Further, what becomes of ideas of justice, rights, power, colonialism, and politics when humans are no longer their sole domain. (23)

But even as political status is granted to the nonhuman, we must be attentive to the extent to which the actions of matter may promote more ethical practices and alternative considerations of justice, power, colonialism, and environmentalism. In the following pages, I draw from the contaminated and reconfigured space of Lake Depue to investigate the emergence of public agency through material rearrangements of space. Specifically, I explore how a heterotopic configuration of industrial waste and festival practices prompts performances from a public that are intended to influence government action to address the contamination of Lake Depue. While the public may be granted greater agency through such material-discursive articulations, a critical lens must be turned, nonetheless, toward these political performances given the way that the heterotopic configuration may disturb familiar notions of environmental ethics and environmental care.

FESTIVAL SPACE AND PUBLIC AGENCY

One spatial configuration that has served as a consistent form of public performance has been the festival, spaces that "range from expositions of high culture, to large-scale popular music extravaganzas, to religious commemorations or thanksgivings, to neighborhood celebrations of a migrant presence, and to statements of alternative sexuality or national pride" (Frost 2016, 570). Even with such variety, festivals offer a consistency in terms of their political and creative potential, which emerges primarily from performances of play (Picard 2016). Frost offers a catalog of such performances, referring to festival spaces as "characterised by risqué reversals of hierarchy, ludic mimicry, flamboyant and celebratory cultural expression, and a sanctioned (or not) overstepping of conventional rules and norms of behavior" (572). While some festivals may enable and even reify existing hegemonic performances and practices of play, other festival spaces enable an alterity and afford the opportunity for publics to adopt new performances and bodies. "The idea that festivals possess a transformative quality is widely accepted by social scientists and has been investigated by researchers approaching the topic from a variety of theoretical angles. Time, space, and social relations are understood to be visibly and affectively transformed through the workings of a festival" (Quinn and Wilks 2017, 36). One commonly referenced example are gay-pride festivals, which are often characterized through the spatiotemporal performances enacted within the festival space that invert and potentially revise social norms. As an other space, gay-pride festivals offer "different forms of articulation of meaning or subjectivity in space" (PlØger 2010, 849). And these different rhetorics and the new meanings and subjectivities such rhetorics construct direct inquiry of festivals into the means by which they "contest existing politics" and the extent to which they "rework historical narratives, revise collective memories, alter the signification of cultural practices and revalue the symbolic capital of the city" (Weller 2013, 2855). The generative potential of festivals emerges through the intra-action of human and nonhuman objects within festival space; as such, these spaces afford an opportunity to consider the influential and agentive capacity of configured, relational matter. While festivals may be seen in terms of a rigid structure in which a consistent set of alternative bodies and practices emerge, we also should not dismiss, following the playfulness afforded by these configurations, the possibility of uncertainty and creativity.

> Although many festival rituals, as well as carnival arts, involve meticulous attention to form and structure, there remains a strong feeling that participation is more than can be conveyed through a measured account of moves, music and costume. The element of risk, of unpredictability, is at the heart of the experience of festivals. (Frost 2016, 572)

To fully account for the transformative potential of these alternative configurations thus entails a move to acknowledge that these spaces and their performances are constituted by more than human intentionality; rather, alterity is made possible by the unexpected movement of and relations among new objects within the assemblage, trajectories and resulting arrangements that may be directed outside of human intentions.

New spaces and subsequently new practices and bodies may emerge by what Massey ([2005] 2015), for example, proposes as happenstance.

> That precisely is one of the truly productive characteristics of material spatiality—its potential for the happenstance juxtaposition of previously unrelated trajectories, that business of walking around a corner and bumping into alterity, of having (somehow, and well or badly) to get on with neighbors who have got here (this block of flats, this neighborhood, this country—this meeting-up) by different routes from you; your being here together is, in that sense, quite uncoordinated. (p. 94)

Massey's explanation of such unpredicted meetings and arrangements may seem to still ground novelty and change in human actions and intentions, yet the same process of seemingly accidental movement and trajectories that in turn foster surprising relations can be applied to nonhuman objects. As Massey further explains, "The nonhuman has its trajectories also and the event of place demands, no less than with the human, a politics of negotiation" (160). The transformative potential inherent in festival space may thus emerge not by way of human interventions and intentional human arrangements of that space but rather through happenstance trajectories of various matter and the resulting configurations, relations, and performances that such novel arrangements and juxtapositions produce. Yet, as Anderson et al. (2012) contend, the acknowledgment of the nonhuman does not mean that individual nonhuman objects can be described as having intentions and motives to change the world for the better, at least through the perspective of a human ethic. Rather, such movements and trajectories of matter are merely products of the continual processes of generating and reconfiguring space and of remaking the world.

Therefore, reconfigurations of space born from the movements of the nonhuman may not always equate to an unequivocal openness and agency that only advances the interests of those previously constrained to act. True to heterotopia, new spatial configurations and the new associations they enable may also afford existing hegemonic practices novel ways to reaffirm their dominance and stability. As a result, the prospect of reconfigurations opening a space for agency prompts a consideration of the following: "How might rethinking causality and agency allow us to consider the dynamic between the durability of assembled orders and their transformation?" (Anderson et al. 2012, 180). As the material-discursive rhetoric used by TECO to constitute the space of Big Bend as a site of sustainability illustrates, the use of the familiar to address incongruity is a challenge to the potential of new agents to generate novel political articulations. Any inquiry must also attend to the possibility that the heterotopic spaces may disturb the practices and ethics that have informed social movements and that have constituted the movements' identities. If material rearrangements afford the potential for change, then such alterations need not be restricted to transforming what may be viewed as the unethical. Heterotopia may also transform the practices, rhetorics, and beliefs that have been associated with the public and ecological good.

THE DIRTY FESTIVAL SPACE OF LAKE DEPUE

The Illinois landscape along Illinois Route 29, a two-lane state highway that runs parallel to the Illinois River, typifies much of the rural American, Midwestern landscape. Traffic jams are nonexistent, and the sparse cars and trucks moving along the road are framed by slightly rolling acres of farm fields. As both Route 29 and the Illinois River begin to make the ninety-degree bend south toward Peoria, Illinois, the traveler comes upon a nondescript metal-green, rectangular sign, pointing to the gray-paved but unlined road that leads to Depue. This pothole-marked county road follows a gradual descent into the Illinois River's floodplain. At the bottom of the slope, the road carries traffic through a sharp right turn to the west, across a dilapidated bridge with posted weight limit signs, and onto heavily cratered Marquette Street, which offers a straight path into the village. The north side of Marquette Street is bordered by an overgrowth of weeds, green vines, and small trees, all kept from encroaching onto the road by way of a six-foot-high, chain-link fence with notable signage: No Trespassing, U.S. Government. Behind the fence line and somewhat hidden by the overgrowth sits a roughly thirty-foot-high mound. Referred to by the locals as the black death, this slag pile contains numerous carcinogenic metals, remnants of the zinc smelter that was once located along this eastern edge of Depue. This fenced site is OU1, the first operational unit of what has come to be called the Depue Group Superfund Site. On the southern side of Marquette Street and directly opposite to the slag pile is the South Ditch or OU4. During the operation of the zinc smelter, the south ditch served to funnel industrial waste water full of carcinogens and metals a few blocks south into holding ponds that bordered Lake Depue. Through flooding and natural erosion and seepage, the metals eventually moved from these holding ponds and South Ditch into Lake Depue, which is now designated as OU5.

Continuing west on Marquette Street, one can follow the fenced Superfund landscape directly into downtown Depue. Like other small towns within the Illinois Valley and other rural villages that have also seen the demise of their industrial base, contemporary downtown Depue consists of very little, primarily the city library with occasional hours and a Casey's General Store. I drove into Depue during the last weekend in July 2018, passed the slag pile, and parked a block south of Casey's along a street lined with mature and healthy-looking trees, offering shelter from the high July sun and placing a canopy of pleasant shade atop old but well-kept residential property. This residential street and the other streets in Depue and the residences that line them have been designated as OU3, a result of the metals emitted into the air during the zinc smelter's operation and that have since settled into the soil of the town's residential landscapes. Lake Depue is situated approximately five residential blocks south of Depue's downtown. Except for the last weekend in July, the lakeshore affords a tranquility that mirrors the quiet of the residential village streets. Surrounded by trees, the lake is well-protected from winds, leading to frequent calm waters, and that pastoral calmness is echoed by the green grass and tree-lined shaded shoreline that borders the southern end of the village. Evenly spaced park benches sit atop the sloped and shaded shoreline and are tucked behind a white picket fence, offering a quiet, restful spot on a summer day. True to the invisible nature of toxins, the Superfund designation and the carcinogenic metals that lay within the residential yards and the lake bed go unnoticed by anyone not informed of their existence, and the presence of the toxins does nothing to visually disrupt the pastoral quality of the lakeshore.

This bucolic scene and the common representation of the natural that the lakeshore invokes have not spurred a rhetoric that has prompted citizen efforts to remediate the lake, however. Instead, the rhetorical capacity to address the toxins emerges by emplacing 30,000 spectators along the lakeshore over the course of three July days as they watch a collection of speedboats emit a high-pitched yowl and, powered by their carbon-fueled motors, race across the tranquil waters. During the July race weekend, the street that runs parallel to the lake becomes populated with food vendors and their food trailers, selling various local foods including tacos, pizza, pastries, and the usual festival food of hamburgers, hot dogs, tenderloins, and elephant ears. Settled alongside the food trailers are a series of small canopies, providing shade for various individuals selling local arts and crafts. The grassy, shade-lined slope of the lakeshore becomes the gathering place for thousands of spectators, some seated on folding lawn chairs, others on blankets placed on the grass. Others stand alongside the lake's banks perched in position to view the boats race around the orange buoys floating atop the lake that mark the wide oval race course. Spaced atop the shaded shoreline are other objects of the weekend festival, including the beer gardens, an entertainment stage, a play area for children, and other food tents and trailers.

The reshaping of the lake, from its tranquil, pastoral quality that marks its existence for fifty-one weeks of the year, to the boisterous, festive gathering that occurs during the July weekend, exhibits the qualities of heterotopic space and the potential transformations that many have aligned with festivals. The festival, while illustrating what some may call a spectacle, plays an important role in constituting a collective, evident in the Facebook posts that mark the continued return by past residents to the festival and the way the boat races serve as a yearly gathering.

> Having grown up in DePue, I have NEVER missed this annual event in my entire life. This little village, with the support of super mayor Eric Bryant and the hard working, loyal members of the fantastic "DePue Mens' Club", offers me and my long time friends an opportunity to enjoy an "annual reunion" every year. You will find the citizens of DePue fun loving and super friendly. A great time to be had by all—long time residents and visitors alike!!!!! (Doll, 2016)

While the festival weekend offers an enactment of community through the various practices of play, the heterotopic juxtaposition of a cultural event and social gathering with toxic metals does not immediately invoke any movement toward addressing the presence of the waste. The transformative capacity of the space only emerges by way of an unexpected configuration of heterogeneous matter, a configuration that occurs by way of the unpredicted movement of various nonhuman actors into the assemblage.

SEDIMENTATION AS A MEANS TO HUMAN AGENCY

Due to La Niña and other weather patterns in 2012, the midwest United States experienced what came to be known as a year without a winter and endured what was recorded as the tenth-warmest year on record (Osborne & Blunden, 2013). The lack of precipitation and the excessive warm temperatures during 2012 had a significant impact on Lake Depue. As a backwater lake within the Illinois River floodplain, Lake Depue receives its water supply largely from seasonal river floods and precipitation. Yet due to the lack of snow during the previous winter, the below-average rainfall during the spring and summer of 2012, and the subsequent absence of river flooding, the lake's water depth heading in the summer of 2012 was far below its average. Moreover, while the seasonal flooding of the Illinois River in previous years had brought the needed water to replenish the lake levels, these same floods had also deposited an excessive amount of sediment within the lake. The increased rate of sediment deposits in Lake Depue can be attributed to several human causes, including the change in agricultural practices within the Illinois River watershed and modifications to the river current and subsequent river flow rates to aid commercial navigation (Demissie, 1996). To ensure the river's use for commercial navigation, locks and dams have been constructed to regulate the water flow and to maintain the needed depth within the shipping channel of the river. "The Illinois River is therefore gradually transforming itself into a narrow river channel in the middle of a wide flood plain without the diversity of small side channels and bottomland lakes" (485). The river's transformation and the resulting excessive sediment are particularly troubling to backwater lakes such as Lake Depue. In the early 1900s, the depth of Lake Depue was, according to residents, anywhere from 10 to 12 feet. A 1976 study contended, however, that the lake bed has risen an average of .59 inch a year due to the excessive sedimentation, steadily lowering the water level of the lake (Lee and Stall 1976, 42).

As the last weekend in July 2012 approached, the lake was too shallow for the boat races to safely occur, presenting the possibility that the yearly boat races and accompanying festival would be canceled. Eleven days before the races were to begin, members of the Depue Men's Club, the organizers of the festival, along with residents of the various communities surrounding Depue converged upon the lake to construct a dam and pump water into the lake in an effort to restore the water to a depth that would allow the boat races to continue (Keyser 2012). The collective group of volunteers along with heavy machinery including bulldozers, backhoes, dump trucks, and farm tractors mobilized at the mouth of the lake to dam the shallow lake waters from further draining into the Illinois River. First, large sandbags were filled with sand and soil and hauled by truck and dumped within the lake's mouth. After two days of sandbagging the mouth and stopping the lake water from flowing into the river, pumps were brought to the site to pull water from the Illinois River into the Lake Depue side of the dam. Through the remaining days, the lake would eventually fill to the depth required to safely allow for the boat races.

By way of the intrusion into festival space of the nonhuman object of sediment and the particular meteorological events of 2012, a collective body and agency emerge that extend beyond just preserving the races and the community festival. Through the happenstance juxtaposition of the various human and other-than-human objects and the resulting configuration, the heroic efforts of the residents become aligned with ridding the lake of the contaminants, with a communal identity emerging that is cast against the neglect and inaction by the state government to rid the lake of the toxic metals. The residents' performance becomes constructed in ways that echo what Buell (1998) references as one of "the constituents of toxic discourse: the moral passion of a battle between David and Goliath" (651). The mayor of Depue, Eric Bryant, affirms the roles adopted in the Superfund struggle: "'We are battling two of the richest companies in the world, and it's kind of like David versus Goliath here,' said Bryant" (Moore 2011, para. 9). While the institutional processes and practices used by the state can be associated with constituting these roles, other nonhuman actors such as the high rate of sedimentation, meteorology, and festival space also play a role in constituting the environmental subjects emplaced within this particular configuration. Through the decades of river flooding, the sediment has added layers of matter atop the lake bed, essentially covering the toxic metals that had migrated decades before from the zinc smelter and the holding ponds. In this regard, the sediment has reconfigured not just the lake and the festival space but also the emplacement and presence of the toxins. The incursion of the deposited sediment into the festival space offers the village some agency and the potential to challenge state officials and the PRP. The potency of the village's challenge emerges not through appropriating the epistemology or rhetoric of techno-science but rather through a performance that will sustain the festival and the boat races. Constructing dams and pumping water may increase the lake's depth, but these efforts do not constrain the continued movement of sediment into the lake and the resulting, perpetual disruption of the festival. For the mayor and village advocates, the requisite act is to dredge the lake to allow the cultural tradition to continue, an act that will also rid the lake bed of the presence of the industrial toxins. But while the sediment's intrusion into festival space has generated agency for the residents, the reconfiguration of the lake by way of the sediment's influence has also enabled an alternative order to emerge, a reality that is constituted by a series of incongruities that in turn foster the need to reconsider what agency, environmental care, and environmentalism mean in a contaminated world.

CONSTITUTING A DIRTY, ENVIRONMENTAL PUBLIC

Years after the group of citizens constructed the dam to maintain the level of the lake, the bed of Lake Depue remains contaminated with metals from the zinc smelter. The subsequent inaction by the PRP since the communal performance in 2012 should not, however, diminish the importance of the acts of the residents nor should the inaction move an understanding of the reconfiguration of the lake to a select moment in time and space. Instead, the reconfigured space brought about by the encroaching sediment and the subsequent performance by the residents enabled a rhetoric and identity of persistence, which continues to serve as a means to generate a collective performance that challenges the PRP practices but that also raises normative concerns pertaining to prevailing notions of environmentalism.

As I wandered around the lakeshore in 2018, I encountered the festive remaking of the place I described earlier, complete with the various food booths, the beer garden, craft tents, and the rows of spectators positioned in their folding chairs. Among all of the objects associated with the festival, however, I encountered no matter within the festival space that immediately generated awareness of the toxic contamination. But while the beer tents and racing boats may not initially appear as objects that configure a space of protest, the yearly persistence and occurrence of the assemblage of these festival objects, nonetheless, invokes a material-discursive rhetoric of protest by the residents and their advocates. As I opened the racing program and scanned past the initial pages of advertisements, I came upon a letter written by Mayor Eric Bryant that serves to introduce readers to the community and to the festival. More than an introductory greeting to visitors and spectators, the letter offers important insight into the way the configuration of the space, including the persistent festival arrangement, the sediment, and the performances of the residents, offers the community agency. The letter opens with the bolded pronouncement: "thirty four years!! and counting." (Bryant 2018, 7). This opening comment refers to the number of consecutive years the races have been held and can be read as an effort to promote the persistence and resolve of the community. In this case, the festive remaking of the lakeshore, especially the fact that the boat races have taken place uninterrupted for thirty-four years, provides a key rhetorical resource for the community, one that affords the opportunity to proclaim that such persistence will continue.

Such a reading is supported in the ensuing set of sentences in which Bryant praises the efforts of a nameless "group of men and women willing to sacrifice time, physical and mental stress and sweat to keep this celebration going every year. With each year the challenges seem to grow but are overcome by a great group of volunteers" (Bryant 2018, 7). The performance initiated in 2012 to combat the sediment and the shallow lake depth has become the source of the recurring emotional and physical sacrifices of the collective. The connection to 2012 is clearly made:

> As I write this letter the water level in Lake Depue is great for races but we are two months away from the races. We must plan for every possible condition the water might be for testing on Wednesday July 25. We have permission to use the inflatable dam we purchased last year for $15000 and a pump for $3500 if we need to pump water. (7)

Bryant draws from the same objects that appeared in 2012 to construct the 2018 festival space. Dams and pumps, in addition to the water, sediment, and the various objects that constitute the boat races, configure the space in which Bryant and the group of residents continue to perform their heroic deeds of persistence. The letter then moves from documenting the efforts to sustain the water level of the lake to the presence of the toxins. The letter, moving to the use of bold type, makes a number of exhortations for communal support and agency. "The battle to save Lake Depue and clean up our village is still being fought!!" (7). Far from passive subjects unable to embark on any action to address the environmental injustice of their village, the residents of Depue are constructed, through the various discursive references to battle, as engaged and capable agents whose effectiveness rests in their persistence. "There is still much work to be done, we must not let up, and we must continue this fight until complete restoration is attained." (7). The letter thus offers an opportunity to understand the way that the festival, the boat races, and the sediment have congealed to promote the restoration of the lake and to prompt the collective subjectivity needed to restore the lake to its precontaminated state.

The rhetoric and collective identity of persistence extends beyond Bryant's letter printed in the festival program and finds itself expressed within the 2014 Village of Depue Comprehensive Plan, which "allows officials and residents to create a vision for short-term and long-term community and economic development" and "explains how the village wants to develop over the next 25 years" (NCICG 2014, 60). The vision of the city conveyed in the plan is rooted in the preservation and cleanup of the lake and the need to address the issues of sedimentation. In discussions of how the lake fits within the futuristic and utopian vision of the village, the plan acknowledges the efforts to maintain the depth of the lake and echoes the persistence of the residents voiced in the festival program. "The village continues to face many challenges, but as history has shown, DePue has always endured" (12). But that endurance is tested by the configuration that has encroached upon and entangled the town. "Unless dredging occurs, the lake will cease to exist and become part of the floodplain of the Illinois River" (39). Should the lake cease to exist, the city would convert into a seemingly undefined, alternative space. "Many residents believe the future of DePue depends largely on remediation of the lake" (11). According to the report, removing the sediment and the toxic metals encased below the river deposits is the only feasible action that can secure the desired vision of the village.

In addition to fostering a subjectivity of persistence and prompting a particular course of action to secure the remediation of the lake, the sedimentation has also enabled the village residents to participate in legislative deliberations in ways previously unavailable. Since 2012, residents of Depue, led primarily by Eric Bryant, have sought permission from various Illinois state governing bodies including the Illinois Department of Natural Resources (IDNR) to install a permanent dam at the lake's mouth. Yet the argument to install a permanent dam extends beyond the weir's capacity to maintain the water level needed to perform the cultural event. Bryant has argued that the dam will need to be in place when dredging occurs to help contain the displacement of the toxicity. "A dam would help keep some of those heavy metals contained within the lake and open up the possibility that the lake could be dredged" (Shute 2013, para. 16). The IDNR, however, has denied the request for a permanent dam, stating the structure would have a negative effect on surrounding wetlands. "'They said it wouldn't be good for fish traveling between the lake and river,' Bryant said, 'We're not happy with that response, but we'll keep trying with the assistance of state official'" (Shute 2016, para. 4). For the IDNR, however, the biota and the practices that constitute the ecology of the lake carry more rhetorical potency than the festival, the racing boats, and subsequently the advocates of the village who seek to permanently dam the lake. This brief exchange introduces the novel incongruities that emerge as environmental justice and environmentalism become at odds. While the incursion of the sediment into festival space has generated a configuration that affords a greater rhetorical capacity for village residents to participate within the discussions of the remediation of the lake, that agency becomes constrained and emplaced within alternative normative concerns as the influence of other actors, especially those typically associated with environmentalism, also gain influence.

IN RESPONSE TO BIOTIC ARTICULATIONS

In the summer of 2018, the PRP conducted a fingernail clam (FNC) study designed to assess the toxicity of Lake Depue, with the results to be included in a baseline ecological risk assessment (BERA). Determining the presence of the toxins in the lake, at least in terms of the institutional practice of risk assessment, requires more than identifying that the industrial waste material exists in the lake bed and in the lake water. The presence of toxicity occurs relationally, a product of the intra-activity of various objects and practices within the lake. As explained in the report documenting the Lake Depue FNC study: "It is commonly recognized that metals concentrations in bulk sediments are typically poor predictors of metal toxicity" (Arcadis 2018, 22). To determine the toxicity of metals, the United States Environmental Protection Agency "has developed guidance for estimating metal toxicity based on the site-specific bioavailable metal fraction" (20). Central to this guidance is the concept of bioavailability, which rests on the assertion that certain configurations of objects, trajectories, and practices generate the toxic agency of metals. Bioavailability can be defined relationally, since "soils and sediments bind chemicals to varying degrees, thus altering their availability to other environmental media (surface water, groundwater, air) and to living organisms (microbes, plants, invertebrates, wildlife, and humans)" (National Research Council 2003, 20). In lay terms, the presence of contaminant material may not be seen as the primary determinant of toxicity; rather, the characteristics of the soils and sediments in which the metals are emplaced determine the extent to which the toxic metals may detrimentally intra-act with the biota in the water and the sediment. Seen in terms of bioavailability, spaces of toxic contamination consist of more than just a collection of inert material objects such as legacy industrial metals; rather, the presence of the waste is constituted by way of the processes and actions invoked by the intra-action among the elements of matter emplaced within such spaces, with the particular composition of sediment playing an influential role in these intra-actions. "Soils and sediments are characterized by intricate associations of biological, chemical, and physical processes that impart functionality in these systems" (122). In this regard, sediment becomes a highly persuasive actor, with a considerable degree of agency that shapes the presence of toxicity and the meaning of a space as contaminated.

To determine the presence of toxicity of Lake Depue, the FNC study investigated the extent to which the metals deposited in the lake bed from the zinc smelter affect the survivability of fingernail clams. Based on their interpretation of the data from the study, the PRP concluded that the primary stressor to the fingernail clams was not zinc or any of the other metals attributable to the smelter but rather ammonia. The IEPA (2019a) contested these results, however: "Dietary exposure to zinc (and other metals) remains a critical unanswered question and Illinois EPA considers this exposure a potential and likely critical stressor which the in-situ toxicity test results have been unable to address" (13). In sum, the IEPA questions the validity of the findings of the PRP that zinc is not bioavailable to the fingernail clams and that its presence cannot be attributed to the deaths of the fingernail clams used in the study. In fact, in response to the initial findings by the PRP-sanctioned agency, the IEPA (2019b) offers a connection between the demise of FNC in Lake Depue with the onset of smelter operations:

> FNCs historically occurred in Lake DePue as evident by the presence of relic shells in sediments deposited prior to the early 1900's. There is a pronounced lack of relic shells in most deeper sediment cores associated with sediment deposited since the early 1900's which appears to be coincident with initiation of smelting operations at the Mineral Point zinc plant. Elevated levels of zinc (and other metals) in surface water, porewater, and sediment may be taken up by FNC through dietary or other exposures to the extent that any internal processes for detoxification or elimination are overwhelmed, thus causing toxicity. (4)

A central practice involved in ingesting zinc occurs through bioturbation, the process used by FNC to burrow and to then feed and breathe. While zinc is present in the surface water of the lake, its presence as a toxin becomes increasingly available through disruptions to the sediment in which the metal is embedded. As such, the IEPA (2019c) offers the following conclusion: "Illinois EPA concludes that total ammonia/un-ionized ammonia and exposures to zinc and other metals (including dietary exposure, which could not be evaluated) are the likely causes of mortality to fingernail clams in Lake Depue" (1–2). The IEPA argues that the presence of zinc presents a dangerous level of toxicity to biota such as fingernail clams, particularly when the toxin becomes available through disruptions to the sediment in which it rests. And the agency also notes that the area of the lake that is the most toxic is the area dredged in the 1980s.

> The FNC test has demonstrated that the central portion of Lake DePue (an area generally coinciding with the previously dredged area) represents an unacceptable ecological risk to FNCs. Site-related stressors are affecting conditions within the central portion of Lake DePue, and the FNC study has shown that these stressors are affecting the survival of FNCs. (IEPA 2019b, 39)

Dredging the central portion of the lake in the 1980s was intended to ensure the continuation of the festival, but the proclamation that this area "represents an unacceptable ecological risk to FNCs" redirects the residents' most recent attempts to dredge the lake to concerns not just with preserving a cultural heritage but also with the effects of disturbing the contaminated sediment and, in turn, enhancing the toxicity of the lake. Deepening the lake by way of removing the deposited sediment may have allowed the festival and cultural event to continue, but the dredging did not remove the contaminants nor did it lessen the harm the toxins posed to biota. Instead, dredging worsened the levels of toxicity by disturbing the sediment and making the contaminants more bioavailable.

CONFRONTING THE CHALLENGES OF AN EMERGENT ENVIRONMENTAL ETHIC

As mentioned, locating ways to enhance a public's agency has been a central concern for those researching sites of environmental injustice. Efforts to advance the legitimacy of alternative knowledges and rhetoric have been cast as a primary means to generate public agency and to promote an alternative and authoritative understanding of living with toxins and waste. The analysis conducted in this paper, however, illustrates the way that matter, through its ability to reconfigure space and to establish novel relations, can also play a role in constituting human subjects with an enhanced performative capacity. From a new materialist perspective, the configuration of Lake Depue thus affords an opportunity to consider rhetoric in terms proposed by Wells et al. (2018): "If change is to be embraced in ontological, affirmative, and hybrid terms, we face new questions. Perhaps rather than inquire what will be most effective, we might ask what will be most potent" (22). Embedded in the assertions by Wells et al. is the familiar new materialist slant that the rearrangement of space, as brought about by matter's vibrancy, generates a novel rhetoric whose potency directs human actors toward some renewed form of ecological care. While the reconfiguration of Lake Depue may have provided residents with agency and a rhetoric that had more potency than any previously available, the reconfiguration also brought about change in terms of the ethics associated with the rhetorical performances of residents and altered prevailing perceptions of environmental justice and environmentalism. As such, the particular configuration of Lake Depue constituted by diverse objects that include sediment, toxic metals, precipitation rates, maritime and agricultural practices, and speedboats offers more than an opportunity to investigate the emergence of new rhetorics. The configuration of Lake Depue also offers a means to explore the ethical considerations of dwelling with waste, specifically in terms of the way the presence of toxins constitutes the need to reorient oneself to emerging, defamiliarized, and dissonant normative concerns.

Perhaps the most obvious point of incongruity rests in the potential of speedboats to act as coconspirators with humans as agents of ecological change. Recreation has long had a contentious relation with environmentalism, evident primarily in the debate as to the most appropriate means for humans to experience natural areas. Human encounters with the natural raise concerns about the extent to which human recreational activities can serve human interests with minimal disruption to the natural ecology (Gottlieb 2005, 63). Hunting has been one of the more noted recreational activities that have come under scrutiny in terms of its relation to ecological care. But as Knezevic (2009) notes, the "greatest threat to all wildlife is not hunting but habitat loss" (13), and the efforts of hunting associations such as Ducks Unlimited "have given us large properties of restored wildlife habitat, replenished wildlife populations, and protected and/or recovered wetland and other wildlife areas throughout North America" (13). Rather than being driven by motives that counter those of traditional environmentalists, hunters taking a strong conservationist approach may enact a similar commitment to ecological care. Besides the weapons associated with hunting, other material objects have played central roles in helping to mediate human relations with the environment and in fostering practices of recreation. While the railroad was responsible for initiating early tourist encounters with and subsequent recreational activities in the national parks of the United States, the automobile became an even more prominent and more democratic object that could provide a greater number of the public with the means to encounter and appreciate the natural. The car provided the transportation to the parks; but the automobile also became an essential part of the interactions with nature through the rise of scenic byways and other roads that promoted contact with the environment (Gottlieb 2005, 66–67). The automobile and the corresponding highways that provided access to the natural parks and natural areas were seen as not only an opportunity to promote the value of these pristine areas but to also advance the "business of scenery" (63).

The skepticism toward the ability of the practices of outdoor recreation to foster an enhanced environmental ethic are noted in DeLuca's (2007) efforts to find some commonality between environmentalism and environmental justice:

> The environmental crisis is the result of a multitude of human practices. If we are going to make any progress in stopping environmental destruction, we are going to have to give up many cultural practices, no matter how much we like them. For example, in the South, from where I am writing, cars and the right to drive them whenever and wherever one wants are considered part of one's cultural heritage—note the devotion to NASCAR. When it comes to saving ecosystems and the planet's health, culture is often

the problem and should not be a trump card used to stop protecting species and ecosystems. (31)

While racing boats, as with racing cars, can be considered as objects and practices unfriendly to the environment, they are, particularly in the configuration of Lake Depue, a prominent object that constitutes the cultural identity of the village. But this cultural object has also become an important actor within matters of environmental injustice and public efforts to eliminate toxins from the lake bed. The particular configuration of the sediment, the racing boats, the geography of the lake, and the human practices to maintain the lake level not only serve to ensure the cultural practice but also provide a performance that is promoted as also protecting biotic species and ecosystems. In this regard, the act of racing across the water becomes reconceived when entangled with the various other-than-human objects that constitute the contemporary space of Lake Depue. As noted, the boats become a crucial material element in constructing the environmental identity of the village residents, and this ecological subject extends beyond efforts to maintain notable cultural practices. The configuration defamiliarizes cultural practices that can be deemed as environmentally destructive. Given the possible cultural and environmental benefits invoked by this particular configuration, questions need to be raised, however, as to why this articulation has struggled to generate political and institutional movement, queries that must include the political influence of the PRP but must also extend beyond the traditional critiques of institutional power common to sites of environmental injustice. Exploring the capacity of the configuration to influence environmental action affords the opportunity to not only understand deliberations within a dirty world but also reveals further ethical complexities and normative dissonance that emerge from the arrangement.

As the IEPA comments referenced earlier convey, dredging the sediment to deepen the lake may enable the continuation of the festival but removing the dirtied muck does not guarantee that the toxic metals from the lake bed will be eliminated or the purity of the lake will be restored. Studies of the lake bed have concluded that the volume of metals increases with the depth of the sediment, a decades-long effect in which a substantial volume of metals has been buried under mud and other biotic debris. Dredging to ensure the appropriate lake depth for the races may disturb the encasement of the metals and may make the toxins more bioavailable, releasing increased volumes of the toxins from the displaced sediment into the lake water. Dredging may thus change the relations among the various objects remaining in the water and in the sediment and produce a more toxic environment. "Disrupting sediment beds to remove contaminated sediments can expose aquatic receptors to otherwise inaccessible contaminants" (Gustavson et al. 2008, 5044). Research also indicates that dredging can have

> short-term adverse effects, including increased contaminant concentrations in the water, increased contaminant concentrations in the tissues of caged fish adjacent to the dredging activity, and short-term increases in tissue contaminant concentrations in other resident biota. (5045)

Subsequently, disturbing the current emplacement of the metals within the sediment, while preserving the cultural event, may further increase toxic threats to the biota inhabiting the lake.

The potential that dredging may enhance rather than eliminate the toxicity in the lake thus raises the need to reconsider the familiar critique that the failure to dredge and the corresponding institutional inaction are products of corporate or state indifference. Institutions charged with the remediation of toxic sites have significant influence over whether dredging or any other means of removing the toxins occur, and those decisions are often informed by numerous factors.

> The decision of whether to remove contaminated sediments is particularly controversial. Superfund operates under a "polluter pays" principle, and of the above options, dredging is typically the most complex and expensive to implement. As such, it is rarely a desirable option for responsible parties at a contaminated site. (Gustavson et al. 2008, 5043)

But if we grant matter a vibrancy and the capacity to politically participate in issues of remediation, the decision to dredge must be considered beyond those interests and intentions of humans. Modern critiques of sites of environmental injustice often associate inaction to political influence, lengthy bureaucratic procedures, and the rigidity of certain institutional practices such as cost/benefit analysis. Yet recognizing the complex entanglements and vibrant matter that characterize posthuman reality introduces other concerns pertaining to the ethics of actions undertaken or even inaction as a means to dwell with dirt. Vibrant matter, as in the case of the toxins of Lake Depue, may suggest that leaving toxins to rest may be the most environmentally ethical path to take if toxicity and its effects on the lake's biota are of ethical importance. Acknowledging the vibrancy of the dirty matter within this particular configuration may not lead to the familiar ethic of care that has informed environmentalism or to the potential that emerges when marginalized publics are granted agency. In other words, the lake will not become remediated and cast as an example of the human capacity to restore damaged lands to the previously pristine state. Nor is dredging feasible to save the boat races and in turn remove the toxins given the proclivity of the buried metals to become more vibrant and move should their current resting place be disturbed. The configuration of Lake Depue surely disrupts the traditional environmental order by offering the public an opportunity to participate in deliberations about the remediation of the lake. But the configuration also disrupts the familiar frames and ethics used to define environmental justice and environmentalism. In the closing section of this chapter, I want to further discuss the influence of the configuration to move such familiar habits of thought and perception.

REGARDING THE ETHICS OF MATERIAL AND RHETORICAL ARRANGEMENTS

One of the central premises informing modern relations to waste is that the dirty matter should be displaced and removed from human encounters. The existence of the boat races within a Superfund site clearly runs counter to these modern perceptions and practices. But the heterotopic quality of the configuration and its resulting capacity to reshape the prospects for dwelling with dirt and the subsequent reorientation of an environmental ethic of care involved in such acts of living become further evident when examining the practices of restoration proposed for Lake Depue and that are being enacted in other areas along the Illinois River. Residents have raised concerns about the tests conducted to determine the extent of the toxicity of Lake Depue, specifically in terms of the depth of the lake bed that has been analyzed. They contend that assessing the toxicity of the lake through sediment samples that do not extend beyond the recently deposited sediment will fail to account for the metals that have since been buried. But beyond providing an avenue to contest the institutional practices, the depth of the metals within the sediment may, in fact, provide a way to dwell with the toxins.

> The dynamic nature of aquatic environments, especially over long periods (tens to hundreds of years), complicates considerations of whether a sediment deposit is stable (i.e., unlikely to be eroded and transported by river channel migration, floods, storms, or other severe events). Many contaminants of concern in sediments are legacy contaminants (e.g., PCBs, DDT, or dioxins and furans) released to the environment decades ago, and they can be buried beneath layers of less contaminated sediments (12, 13). Contaminants buried below the biologically active zone—the upper layers of sediment where organisms live or interact—are neither accessible nor available to sediment- and water-dwelling organisms. (Gustavson et al. 2008, 5043–5044)

Subsequently, toxins that are buried under a sufficient amount of sediment may pose little harm to entangled biota, while disturbing the sediment through dredging, for example, may result in an increased level and availability of toxins. The influence of the association between the amount of sediment and the toxic metals on the institutions overseeing the remediation of Lake Depue becomes evident in a 2004 report funded by the United States Environmental Protective Agency Superfund Redevelopment Initiative. The report concluded that "despite the significant role that boat races play in the culture and history of Depue it may not be feasible to remove contaminants in the lake sediment" (E^2 Inc. and D.I.R.T Studio 2004, 13). The challenge to enact the familiar practice of dredging and removing the toxins is due to the significant amount of sedimentation that now characterizes the Illinois River valley and that greatly affects the existence of backwater lakes such as Lake Depue.

> The most significant impact of soil erosion in the Illinois River basin is the high rate of sedimentation in the bottomland lakes of the Illinois River valley. These lakes are remnants of a much larger glacial river system that once occupied the Illinois River valley, and they provide important ecological and recreational functions. The bottomland lakes have lost on the average more than 72% of their capacities, with some of them now completely filled with sediment. Although there is a great interest in restoring or saving some of the lakes along the river, this has been a very difficult task, because of the high rates of sediment delivery into the Illinois River valley. (Demissie 1996, 483)

Dredging the lake to not only eliminate the buried toxins but to also ensure the depth needed to sustain the boat races may not be feasible, and not just because of the potential to disrupt the legacy toxins within the buried sediment. Dredging the lake will not eliminate the future sediment making its way downstream from continuing to encroach upon and reshape the lake. Even if the buried toxins could be removed, excessive rates of sedimentation would continue to fill the lake and continue to threaten the existence of the cultural event. Living with the toxic waste thus means acknowledging the other objects and practices threatening the lake's existence, objects that also constrain the public's agency.

These additional actors, as the E^2 and D.I.R.T Studio report concedes, have significant influence in not only reshaping the lake but in fostering novel human activities that some may conceive as less environmentally threatening than carbon-fueled boat races. While the report acknowledges that "the Village of Depue will suffer a significant loss of cultural heritage" should the races no longer take place, "dredging is not necessary for the lake to provide significant benefits to the community as wildlife habitat and an eco-tourism attraction" (13). Instead of dredging the lake to ensure the continuation of the festival and to address the presence of the toxins, the report draws on the material-discursive rhetoric of environmentalism and the influence of the material configuration to promote a new vision in which the lake is transformed into a wetland, which will not only contain and eventually eliminate the contamination but also provide an opportunity for a more familiar form of environmental care. Moreover, the ecological transformation would provide benefits beyond the residents and festivalgoers: "This habitat enhancement would help ducks, geese, all varieties of shorebirds, egrets, herons, eagles, bitterns, beavers, otters, muskrat, minks, weasels, and raccoons, as well as other animals. Additionally, aquatic vertebrates and invertebrates would benefit" (33). The report's rhetoric, which takes up the familiar tropes of environmentalism by invoking images of a restored Eden, raises some fundamental concerns with respect to an ethic of care in a contaminated world, however. The transformation of the lake to a wetland may be seen as an ethical act in relation to the nonhuman objects that may populate such a space. But some consideration must be given to the loss of the cultural festival, especially given the decline of the village resulting from the deterioration of its industrial and economic base. Further, the current configuration of the lake offers a suitable habitat for wildlife, and the proposed wetland would offer an environment no less toxic than the current configuration. In both spaces, the toxins still exist, encased within the sediments of the lake bed.

The E^2 and D.I.R.T Studio plan (2004), in many ways, echoes the rhetoric common to other spaces of remediation, including the construction of the STA's to remediate the contaminated waters of Lake Okeechobee mentioned in a previous chapter. As the plan advances:

> Many of the contaminants found on the DePue/New Jersey Zinc/Mobil Chemical Corporation Superfund site, such as arsenic, cadmium, lead, zinc, copper, cyanide, iron, and selenium can be remediated using natural or constructed wetlands. The ability of wetlands to cleanse contaminants from soil and surface water was discovered in the early 1990s. Since then, several contaminated sites around the country have been successfully remediated through the creation and restoration of wetlands. (33)

Yet as the efforts to direct and treat tainted water from Lake Okeechobee by way of the STA's demonstrate, such promises need to be tempered with the uncertainty that such spaces are able to perform as promised given the likelihood of unpredictable disturbances. And since the source of Lake Depue's water and any proposed wetland is the Illinois River, the potential of the transformed wetland to address the toxins from the zinc smelter must also be recognized as being entangled within the toxic legacy of the river. The lake is a space shaped by upriver industrial and engineering practices, including the redirection of water that once flowed to Lake Michigan but now flows into the Illinois River.

> On January 1, 1900, the Chicago Sanitary and Ship canal opened. This canal connected the Des Plaines and Illinois Rivers to Lake Michigan and as a result gave the City [sic] of Chicago a means of flushing untreated domestic sewage and industrial wastes away from Lake Michigan into the Illinois River system by diverting water from Lake Michigan into the Illinois River. (USACE 2007, I-22)

Altering the movement of water away from Lake Michigan has redirected the waste from the metropolitan area of Chicago, to include industrial waste from facilities now populating the Sanitary and Ship Canal, into the Illinois River, creating a waterway tainted with mercury, PCBs, and various other toxins. And the seasonal floods that are essential to resupplying Lake Depue with water carry these various contaminants into the lake. As the IEPA (2016) notes: "A fish advisory has been issued for Lake DePue for carp, white bass, and channel catfish due to detections of PCBs in fish tissue samples. In addition, there is a state-wide fish advisory for methyl mercury in predator fish" (para. 18). Lake Depue is contaminated not just from the industry that was located in the village; the lake is also dirtied by other toxins that make their way down river and eventually, through the seasonal floods, into the lake waters. Yet in the E^2 and D.I.R.T. Studio plan and in the performances by the residents, the only attention is given to the toxicity of the lake with respect to the matter that has derived from the zinc smelter. In either case of remediation, whether it be in the dredging of the lake sediment or in the transformation of the lake into a wetland, the water will still remain contaminated with the excessive levels of mercury and PCBs.

The disarticulation of these river toxins as an element within efforts to remediate the lake merits consideration within an ethic of dwelling within contemporary spaces of waste. Giddens (1990) characterizes modernity by the continual presence of toxins and other risks:

> The sheer number of serious risks in respect of socialized nature is quite daunting: radiation from major accidents at nuclear-power stations or from nuclear waste; chemical pollution of the seas sufficient to destroy phytoplankton that renews much of the oxygen in the atmosphere; a "greenhouse effect" deriving from the atmospheric pollutants which attack to the ozone layer, melting part of the ice caps and flooding vast areas; the destruction of large areas of the rain forest which are a basic source of renewable oxygen; and the exhaustion of millions of acres of topsoil as a result of widespread use of artificial fertilizers. (127)

Giddens then explains the affects and perceptions common to human subjects created by the material existence of such an abundance of risk: "One is the numbing feeling, almost of boredom, which such a list is likely to induce in the reader—a phenomenon which relates to the sixth point in the risk profile, the fact that awareness of many generalized kinds of risk is now widespread among the population at large" (127–128). Rather than exhibiting a numbness or a willful blindness to these river toxins, the residents of Depue may instead adopt a position Giddens refers to as pragmatic acceptance, manifest in "the belief that much that goes on in the modern world is outside anyone's control, so that temporary gains are all that can be planned or hoped for (135). Giddens assigns such a bodily state to laypeople, asserting that it is "not so much a withdrawal from the outside world as a pragmatic participation which maintains a focus on day-to-day problems and tasks" (135). The capacity to save the festival and, in turn, to remove the toxins from Lake Depue results from the perception of having some degree of agency and influence upon the world, a subjectivity that becomes manifest in the construction of the dam to raise the lake water level and that becomes furthered through the various material-discursive performances to save the races and to promote the dredging of the lake. The configuration, however, does not seem to invoke similar performances that generate attention or action toward the other toxic matter that also constitutes the dirtied lake. Even environmentalists, who have promoted and taken various actions to address the levels of mercury in the river, nonetheless, find themselves having to accept that the river will remain, to some degree, dirtied. Novel human action and agency are possible through the emergence of new configurations, but that agency may clearly be constrained by other objects within the entanglement. In a contaminated world, such reconfigurations may not offer the means toward a renewed ethic of care and may, in fact, force us to consider the limits of drawing from familiar ethical frames to perceive these emergent realties and accept the possibilities and constraints of pragmatic participation within a dirtied world. In other words, inaction may not result from will blindness but rather because of an incapacity to act.

One additional practice aimed at addressing the demise of the backwater lakes along the Illinois River further clarifies the need to reconsider ethical practices and the potential of actors in contaminated spaces. In 2000, The Nature Conservancy purchased a parcel of land along the western shore of the Illinois River and roughly 100 miles south of Lake Depue. Previously committed to agriculture, the purchased land, now called the Emiquon Preserve, has been transformed into a wildlife refuge, as flora previous to the agricultural intrusion were replanted and water flow was redirected to convert the space to a viable wetland ecology. Central to the existence of the preserve, however, is the technology used to manage the flow of water. The preserve is not a place where control has been returned to nature and where the present ecology can be said to exist in its natural, undisturbed state. Instead, human actors and specific technologies are used to construct the space and the ecology that constitutes it.

> Water management means the drainage district, in consultation with the Conservancy and partners, will actively direct the amount and timing of water flow out of or into Emiquon by opening or closing the gates or using pumps to recreate natural water level fluctuations needed to sustain the high-quality habitats in Emiquon. (The Nature Conservancy 2021a, para. 11)

A central component to constructing this space and to managing the water that flows between the river and the backwater wetland is the gated pathway referred to as the Ahsapa Water Control Structure.

> The structure makes it possible to isolate the preserve from the river when desirable to avoid heavy loads of sediment, nutrients, contaminants and/or invasive species. By providing drainage when desirable, the structure will help manage sediments in Emiquon. Long before there is a sediment problem, the gates and pumps could be used to draw water levels down, thereby drying and compacting the sediments. During the dry period, sediments could also be physically removed and used to strengthen the levee or for other beneficial purposes. Additionally, Emiquon could benefit from periodic supplements of sediment and associated nutrients provided by a controlled influx of river water. (para. 15)

Similarities exist between the rhetoric used to describe Emiquon and the discourse found in the E^2 and D.I.R.T report describing the proposed wetland that would contain the toxins in Lake Depue. Echoing the wide-ranging, environmental benefits brought about by the transformation of Lake Depue into a wetland, Emiquon also offers an opportunity to restore the vibrancy and heterogeneity that once characterized the backwater ecologies of the Illinois River while also providing regional, economic benefits: "With a wide-spread reputation for great fishing, hunting and birding, Emiquon attracts sportsmen and women from throughout the state, bringing more business to local sporting goods shops and the restaurants where these visitors eat" (The Nature Conservancy 2021b, para. 13). Restoration that favors nonhuman actors and recreation activities, specifically those that are deemed more environmentally responsible than activities associated with carbon-based speedboats, become important elements to how both spaces become articulated and perceived.

But while the vision of Lake Depue offered in the E^2 and D.I.R.T report and the goals of Emiquon as expressed by The Nature Conservancy may seem similar, clear differences emerge that reveal the ethical limits of converting Lake Depue into a wetland. In addition to generating the wetland ecology, the water management practices at Emiquon are designed to also address the various stresses impacting the Illinois River, including invasive species, excessive sediment, and contamination, in order to prevent, as much as possible, these stressors from impacting the welfare of the wetlands. Emiquon acknowledges the clear and immediate relationality that exists between the wetland and the Illinois River. And The Nature Conservancy readily acknowledges that in the act of transformation and restoration, human interjection is crucial to its success, and given the incursion of what are cast as numerous maleficent actors, such human action is even more important. The E^2 and D.I.R.T report, in contrast, neglects the relationality between the river and the lake and omits any reference to the need for continual human intervention, suggesting that simply letting the lake revert to natural conditions will transform the body of water back to its previously pristine condition. Burying the toxic metals under the sediment may be the most feasible means to address the contamination, given the questionable efficacy of dredging and removing the dirty matter; but this action neglects the need to attend to the toxic contaminants that exist within the sediment and the dirty water flowing into the converted wetland from the Illinois River. Moreover, the report neglects the continual incursion and disruption that the ever-present and increasing amount of sediment will produce. The wetland transformation of Lake Depue entails no actions to manage future toxic, sediment loads.

But arguments by residents to dredge the lake must also be reviewed within this same relational, ethical context. Neither the dredging proposed by the residents or the restoration of the lake to a wetland will lead to a cleansed and pure Edenic and utopian ecology. Dredging the lake will not remove the toxic sediments in the lake bed, and, as the past dredging shows, may, in fact, generate a more toxic configuration. Additionally, the residents' proposal also omits the continual incursions and disturbances that the river sediment will provoke. Without specific actions to address the sediment loads, dredging will, at some future point, need to take place again. As such, the discursive articulation generated by the residents, a rhetorical association made possible by the material configuration of the sediment, toxins, lake bed, boat races, and precipitation events, must be ethically challenged. The happenstance configuration that has assembled the toxins, the depth of the lake, and the boat races may have provided the residents with an agency previously unrealized, but the quality of the relationality among these actors and practices as a form of social action raises questions as to the ethics of such a rhetoric and the resulting mode of environmental advocacy. The relation between saving the races and removing the toxins as mediated by the act of dredging the sediment becomes tenuous when considering the influence of the various nonhuman actors within the configuration and the constraints placed upon human practices. The articulations that emerge in Lake Depue reveal that new configurations may indeed provoke new rhetorics, but the rhetorics that emerge and as constituted by novel relations must be considered through an ethics that emerges from these relations.

Conclusion

ARTICULATING RESPONSIBILITY IN A DIRTIED WORLD

As the title of this book suggests, this project has centered on a search for ways to engage with the novel and dissonant spaces resulting from an increasingly, dirty world. While some specific conclusions may be drawn from the heterotopia that have been discussed, I hope, as with other lines of research, that the exploration of the spaces offered in this book also prompts continued interest and inquiry. I hope to have illustrated through the analysis of the contemporary configurations that constitute Lake Depue, Weldon Spring, and central and coastal Florida that the familiar understandings and responses to the presence of waste and to the disturbances that such encounters provoke may no longer appropriately accommodate the emplacement of humans within dirtied and dissonant spaces. As waste continues to populate the planet and creeps into those spaces once perceived as absent of dirty matter, we, nevertheless, continue to rely on the familiar practices of attempting to (re)move waste, further instilling the belief that the matter is absent from our daily lives. We continue calls for lower emissions and the elimination of the contaminants that dirty our air, land, water, and bodies, but these efforts do not appropriately equip us to meet the daily tasks of ethically living with the soiled spaces we have created and inhabit, the persistent presence of waste, and waste's capacity to reshape worldly spaces.

We refer to dirtied spaces like responsible parents. We point to these spaces, remarking that this is what happens when we engage in capitalism's excesses; but we fail to fully appropriately acknowledge the implications and responsibilities of living with the spaces created by this behavior. We turn our attention to how these spoiled sites may prompt us to refrain from the wickedness of consumerism, as we turn over the care of and responsibility for dirtied spaces to someone else. While such earthly and ecological parenting is important, it, nonetheless, neglects the persistence and vibrancy of the dirtied configurations and the potential of these heterotopia to prompt public responsibility toward their care. The familiar environmental rhetoric may generate important public action, notably in the awareness to one's complicity in the production of the dirt. But the heterotopia discussed in this book offer the opportunity to move beyond the guilt and shame of the publics' complicity so as to turn attention to the publics' responsiveness to and responsibility for contemporary, dirtied configurations. Questions thus emerge as to how we understand and respond to the dirty configurations that constitute dissonance with familiar forms of environmentalism, such as the possibilities that emerge when speedboats may act as environmental advocates, when waste becomes a viable tourist object to promote an understanding of the contemporary ecology and one's responsibilities for it, and when one's responsiveness and responsibility to the presence of dirt become materially and discursively articulated in novel and unexpected ways.

The dirtied ecologies discussed in this book invoke the need to base ecological care on a capacity to understand one's performance in the dirtied world as a product of relations that emerge, hold together, and fray in novel and unpredictable ways. Associations among waste, humans, and nonhumans may no longer be easily tied to downwind or downstream logics of cause and effect, production and consumption, and human intentions but rather constituted through the unpredicted movements and manners of dispersed and influential vibrant material. As the familiar no longer becomes tenable, an ethic of care then must emerge from the creative, from the possibilities afforded by novel configurations but also through the imaginative potential of the human actor whose ethical performances hinge on the proficiency to interpret and respond to the novel potential such spaces provide. Such human potential must be understood through the responsibilities humans have to these new ecologies and the responsiveness these ecologies afford. The AMD & ART project discussed in this book, for example, offered a space for a creative response to the presence of waste, but humans failed to live up to their responsibilities to appropriately and affectively care for this new space. The novel configuration of Weldon Spring also presents the opportunity for a creative response to waste's presence and a reorientation to public responsibilities to such dirtied sites. Through their encounter with waste, publics are given the capacity to comprehend the relationality among various human and nonhuman actors who constitute this novel ecology, an understanding that may stimulate the visiting public to recognize its responsibilities as a participant within the configuration and as a participant within the larger, geographically dispersed configurations that constitute the dirty world. The entanglement that comprises central and coastal Florida, a space that extends beyond the borders of the Manatee Viewing Center to also encompass the state's natural springs, residential developments, the Everglades, and the coastal biota and human populations, illustrates that efforts to develop a responsiveness to contemporary, ecological care become confounded by a configuration's wide range of influential and accountable participants. An ethic of care in a dirtied world thus becomes associated with the capacity to generate meaningful articulations and associations among seemingly discordant and geographically separate human and nonhuman actors. And a contemporary ethic of care must contend with the constraints placed upon human actions and intentions. Such limits are clearly evident in the novel configuration of Lake Depue, in both the citizen efforts to associate the boat races with removing the toxins from the lake and the institutional desires to allow the lake to become transformed into a wetland. Both actions hold fast to the modern and familiar promise of human potential to recreate a greener world; yet both performances hold a measure of deceptiveness, given their dismissal of the persistent presence of dirty matter.

The willful ignorance that may constitute contemporary ecological subjects' relation to dirty emissions and waste, therefore, may not just be attributed to dominant, capitalist discourses that make invisible the dirt produced through the modern, economic machine; the publics' willful ignorance may also be a product of the familiar tropes of environmentalism that make invisible the challenges, opportunities, and responsibilities of living and playing within all the spaces of perpetual dirt. Environmental justice advocates have performed valued work to raise public awareness of the dirty spaces populated by the marginalized. And while waste remains unequally distributed, its growing presence in various public spaces presents the opportunity to further extend public engagement as to how humans may respond to dirty configurations and how such performances also align with matters of equity.

Environmentalists have a role in fostering such proficiency and in turn moving publics toward a new relation and responsibility to the dirtied, vibrant ecologies that constitute contemporary spaces, a proficiency rooted in fostering understandings and responses to novel configurations and relations. The acceptance of waste's presence and of dirtied spaces does not correspond to yielding to capitalism's detrimental effects nor conceding that modern human reasoning and inventiveness can cure all industrial ills. The opportunities afforded by the material-discursive articulations of contemporary heterotopia, instead, call upon publics to be responsive and responsible to the impacts of industrialism. Moreover, the constraints and challenges evident in the various performances discussed in this book clearly show that perceptions of human mastery are no longer fitting. The proficiency to understand and respond to these spaces of waste holds the potential to more clearly present the need for humans to become ethically adaptable and imaginative to their responsibilities while recognizing the constraints on these efforts, a proficiency that will likewise be needed given the ecological changes brought about by climate change. It is crucial to build the institutional capacity for resilience, as Benson and Craig posit; but it is crucial to reorient environmentalism's reach so as to engage publics with the requisite proficiency to become ethically equipped to respond to and responsibly perform within the dirtied spaces that mark our contemporary ecology.

Bibliography

Ackerman, John. 2003. "The Space for Rhetoric in Everyday Life." In *Towards a Rhetoric of Everyday Life: New Directions in Research on Writing, Text, and Discourse*, edited by P. Martin Nystrand and John Duffy, 84–117. Madison: University of Wisconsin Press.
Alaimo, Stacy. 2012. "Sustainable This, Sustainable That: New Materialisms, Posthumanism, and Unknown Futures." *PMLA* 127 (3): 558–64: https://doi.org/10.1632/pmla.2012.127.3.558.
AMD & ART. 2016a. "AMD&ART, The Project: 1994–2005." https://amdandart.info/.
———. 2016b. "The Wetlands: History." https://amdandart.info/tour_wetlands1.html.
Anderson, Ben, Matthew Kearnes, Colin McFarlane, and Dan Swanton. 2012. "On Assemblages and Geography." *Dialogues in Human Geography* 2 (2): 171–89. https://doi.org/10.1177/2043820612449261.
Arcadis. 2018. *Technical Memorandum: In-situ Caged Fingernail Clam Study Results. Operable Unit 5, Depue Lake, Depue Site, Illinois*. IEPA Document Explorer, November 27, 2018. https://external.epa.illinois.gov/DocumentExplorer/Documents/Index/170000002878.
Audubon Florida. 2016. "The Everglades Agricultural Area Reservoir Project: An Urgent Call to Action." https://fl.audubon.org/sites/default/files/audubon_eaa_reservoir_may2016.pdf.
Badruzzaman, Mohammad, Jimena Pinzon, Joan Oppenheimer, and Joseph G. Jacangelo. 2012. "Sources of Nutrients Impacting Surface Waters in Florida: A Review. *Journal of Environmental Management* 109: 80–92. https://doi.org/10.1016/j.jenvman.2012.04.040.
Barad, Karen. 2007. *Meeting the Universe Halfway: Quantum Physics and the Entanglement of Matter and Meaning*. Durham: Duke University Press.

Barker, Jacob. 2015. "New Documentary Walks Viewers Through St. Louis Nuclear History." *St. Louis Post-Dispatch.* July 17, 2015. https://www.stltoday.com/business/local/new-documentary-walks-viewers-through-st-louis-nuclear-history/article_b04f9995-a80a-52bf-99ac-798ffd73142b.html.

Barnett, Joshua Trey. 2015. "Toxic Portraits: Resisting Multiple Invisibilities in the Environmental Justice Movement." *Quarterly Journal of Speech* 101 (2): 405–25. https://doi.org/10.1080/00335630.2015.1005121.

Baumlin, James S. 2020. "From Postmodernism to Posthumanism: Theorizing Ethos in an Age of Pandemic." *Humanities* 9 (46): 1–25. https://doi.org/10.3390/h9020046.

Beck, Eckardt C. 1979. "The Love Canal Tragedy." *EPA Journal, The Economy and the Environment* 5 (1): 16–19. https://www.epaalumni.org/history/journals/?searchterm=love+canal&searchtype=0.

Beck, Ulrich. (1986) 2011. *Risk Society: Towards a New Modernity,* translated by Mark Ritter. London: Sage.

———. 1998. "The Challenges of a World Risk Society." *Korea Journal* 38 (4): 196–205.

Bennett, Jane. 2010. *Vibrant Matter: A Political Ecology of Things.* Durham: Duke University Press.

Bergthaller, Hannes. 2014. "Limits of Agency: Notes on the Material Turn from a Systems-Theoretical Perspective." In *Material Ecocriticism,* edited by Serenella Iovino & Serpil Oppermann, 37–50. Bloomington: Indiana University Press.

Benson, Melinda Harm, and Robin Kundis Craig. 2017. *The End of Sustainability: Resilience and the Future of Environmental Governance in the Anthropocene.* Lawrence: University Press of Kansas.

Bitzer, Lloyd F. 1968. "The Rhetorical Situation." *Philosophy & Rhetoric* 1 (1): 1–14.

Blair, Carole. 1999. "Contemporary U.S. Memorial Sites as Exemplars of Rhetoric's Materiality." In *Rhetorical Bodies,* edited by Jack Selzer and Sharon Crowley, 16–57. Madison: University of Wisconsin Press.

Blair, Carole, Greg Dickinson, and Brian L. Ott. 2010. "Introduction: Rhetoric/Memory/Place." In *Places of Public Memory: The Rhetoric of Museums and Memorials,* edited by Greg Dickinson, Carole Blair, and Brian L. Ott, 1–54. Tuscaloosa: University of Alabama Press.

Boissoneault, Lorraine. 2018. "The Deadly Denora Smog of 1948 Spurred Environmental Protection—But Have We Forgotten the Lesson?" *Smithsonian Magazine,* October 26, 2018. https://www.smithsonianmag.com/history/deadly-donora-smog-1948-spurred-environmental-protection-have-we-forgotten-lesson-180970533/.

Boley, B. Bynum and Gary T. Green. 2016. "Ecotourism and Natural Resource Conservation: The 'Potential' for a Sustainable Symbiotic Relationship." *Journal of Ecotourism* 15 (1): 36–50. https://10.1080/14724049.2015.1094080.

Bowers, Tom. 2018. "Heterotopia and Actor-Network Theory: Visualizing the Normalization of Remediated Landscapes." *Space and Culture* 21 (3): 233–46.

———. 2020. "Speculations on the Political Agency of Public Matter." *Journal of Multicultural Discourses.* https://doi.org/10.1080/17447143.2020.1857766.

Braidotti, Rosi. 2013. *The Posthuman.* Cambridge: Polity Press.

———. 2019. *Posthuman Knowledge.* Cambridge: Polity Press.

Brown, Kate. 2019. "Learning to Read the Great Chernobyl Acceleration: Literacy in the More-Than Human Landscapes." *Current Anthropology* 60 (20): S198–S208. https://doi.org/10.1086/702901.
Bryant, Eric. 2018. "Letter to Mr. Steve Solorio." *Lake Depue '18 Pro National Championship Boat Races.*
Buell, Frederick. 2003. *From Apocalypse to Way of Life: Environmental Crisis in the American Century.* New York: Routledge
Buell, Lawrence. 1998. "Toxic Discourse." *Critical Inquiry* 24 (3): 639–65.
Bullard, Robert D. 2018. *Dumping in Dixie: Race, Class and Environmental Quality.* 3rd ed. New York: Routledge.
Bureau County Tourism Committee. 2021. "Experience Bureau County Illinois: 2021 Visitors Information Guide." https://www.bureaucounty-il.com/index.
Callahan, Jr., Richard J., Kathryn Lofton, and Chad E. Seales. 2010. "Allegories of Progress: Industrial Religion in the United States." *Journal of the American Academy of Religion* 78 (1): 1–39. https://doi.org/10.1093/jaarel/lfp076.
Capece, John C., Kenneth L. Campbell, Patrick J. Bohlen, Donald A. Graetz, and Kenneth M. Portier. 2007. "Soil Phosphorus, Cattle Stocking Rates, and Water Quality in Subtropical Pastures in Florida USA." *Rangeland Ecology & Management* 60 (1): 19–30. https://doi.org/10.2111/05-072R1.1.
Carr, John, and Tema Milstein. 2021. "'See Nothing but Beauty:' The Shared Work of Making Anthropogenic Destruction Invisible to the Human Eye." *Geoforum* 122: 183–192. https://doi.org/10.1016/j.geoforum.2021.04.013.
City of Deltona. n.d. *City History*, accessed June 20, 2020. https://www.deltonafl.gov/home/pages/city-history.
Connolly, William E. 2010. "Materialities of Experience." In *New Materialisms: Ontology, Agency, and Politics*, edited by Diana Coole & Samantha Frost, 178–200. Durham: Duke University Press.
Coole, Diana and Samantha Frost. 2010. "Introducing the New Materialisms." In *New Materialisms: Ontology, Agency, and Politics*, edited by Diana Coole and Samantha Frost, 1–43. Durham: Duke University Press.
Cortez, Marisol. 2012. "Time Out of Mind: The Animation of Obsolescence in 'The Brave Little Toaster.'" In *Histories of the Dustheap: Waste, Material Cultures, Social Justice*, edited by Stephanie Foote and Elizabeth Mazzolini, 227–251. Cambridge: MIT Press.
Cowardin, Lewis M., Virginia Carter, Francis C. Golet, and Edward T. LaRoe. 1979. *Estuarine System. Classification of Wetlands and Deepwater Habitats of the United States.* U.S. Department of the Interior, Fish and Wildlife Service, Washington, D.C. Jamestown, ND: Northern Prairie Wildlife Research Center Online. https://www.fws.gov/wetlands/documents/classwet/index.html-contents (Version 04DEC1998).
Cox, Robert. 2013. *Environmental Communication and the Public Sphere*, 3rd Edition. Los Angeles: Sage.
Cronon, William. 1996a. "Introduction: In Search of Nature." In *Uncommon Ground: Rethinking the Human Place in Nature*, edited by William Cronon, 23–56. New York: W.W. Norton & Company.

———. 1996b. "The Trouble With Wilderness: Or, Getting Back to the Wrong Nature." In *Uncommon Ground: Rethinking the Human Place in Nature*, edited by William Cronon, 69–90. New York: W.W. Norton & Company.

Cudworth, Erica, and Stephen Hobden. 2018. *The Emancipatory Project of Posthumanism*. New York: Routledge.

Davis, Reade, and Laura Zanotti. 2014. "Introduction: Hybrid Landscapes: Science, Conservation, and the Production of Nature." *Anthropological Review* 87 (3): 601–611. https://doi.org/10.1353/anq.2014.0032.

De Coverly, Edd, Pierre McDonagh, Lisa O'Malley, and Maurice Patterson. 2008. "Hidden Mountain: The Social Avoidance of Waste." *Journal of Macromarketing* 28 (3): 289–303. https://doi.org/10.1177/0276146708320442.

DeLuca, Kevin Michael. 1999a. *Image Politics: The New Rhetoric of Environmental Activism*. New York: The Guilford Press.

———. 1999b. "Articulation Theory: A Discursive Grounding for Rhetorical Practice." *Philosophy & Rhetoric* 32 (4): 334–48. https://www.jstor.org/stable/40238046.

———. 2007. "A Wilderness Environmentalism Manifesto: Contesting the Infinite Self-Absorption of Humans." In *Environmental Justice and Environmentalism: The Social Justice Challenge to the Environmental Movement*, edited by Ronald Sandler and Phaedra C. Pezzullo, 27–55. Cambridge: MIT Press.

DeLuca, Kevin Michael, and Anne Teresa Demo. 2000. "Imaging Nature: Watkins, Yosemite, and the Birth of Environmentalism." *Critical Studies in Communication* 17 (3): 241–60. https://doi.org/10.1080/15295030009388395.

Demissie, Misganaw. 1996. "Patterns of Erosion and Sedimentation in the Illinois River Basin." *Erosion and Sediment Yield: Global and Regional Perspectives (Proceedings of the Exeter Symposium, July 1996)*: 483–90. https://www.researchgate.net/publication/242261832_Patterns_of_erosion_and_sedimentation_in_the_Illinois_River_basin/link/55254d370cf2561f2ac28958/download.

Doll, Rick. 2016. "Rick Doll Reviewed Depue Boat Races." Depue Boat Races Facebook, July 9, 2016. https://www.facebook.com/pg/depueboatraces/reviews/?referrer=page_recommendations_see_all&ref=page_internal.

Donaldson, L. n.d. "General Information for Basin Management Action Plans-BMAP." Florida Department of Environmental Protection, accessed June 20, 2020. https://floridadep.gov/comm/comm/documents/general-information-basin-management-action-plans-bmap-0.

Donaldson, L. 2019. *Blue-Green Algae Task Force Draft Consensus Document #1*. October 11, 2019. https://floridadep.gov/comm/comm/documents/blue-green-algae-task-force-draft-consensus-document-1.

Douglas, Mary. (1966) 2002. *Purity and Danger: An Analysis of Concept of Pollution and Taboo*. London: Routledge.

E^2 Inc. and D.I.R.T. Studio. 2004. *Depue/New Jersey Zinc/Mobil Chemical Corporation Superfund Site Project Report 2004*. Funded by United States Environmental Protection Agency Superfund Redevelopment Initiative (SRI) http://semspub.epa.gov/work/05/633176.pdf.

Edbauer, Jenny. 2005. "Unframing Models of Public Distribution: From Rhetorical Situation to Rhetorical Ecologies." *Rhetorical Society Quarterly* 35 (4): 5–24. http://www.jstor.org/stable/40232607.

Eden, Sally, Sylvia M. Tunstall, and Susan Tapsell. 1999. "Environmental Restoration: Environmental Management or Environmental Threat?" *Area* 31 (2): 151–59. https://www.jstor.org/stable/20003970.

Edensor, Tim. 2005. "Waste Matter: The Debris of Industrial Ruins and the Disordering of the Material World." *Journal of Material Culture* 10 (3): 311–32. https://doi.org/10.1177/1359183505057346.

Endres, Danielle. 2018. "The Most Nuclear-Bombed Place: Ecological Implications of the U.S. Nuclear Testing Program." In *Tracing Rhetoric and Material Life: Ecological Approaches*, edited by Bridie McGreavy, Justine Wells, George. F. McHendry. Jr., and Samantha Senda-Cook, 253–87. Cham: Palgrave Macmillan.

Endres, Danielle, and Samantha Senda-Cook. 2011. "Location Matters: The Rhetoric of Place in Protest." *Quarterly Journal of Speech* 97 (3): 257–82. https://doi.org/10.1080/00335630.2011.585167.

Ewalt, Joshua P. 2016. "The Agency of the Spatial." *Women's Studies in Communication* 39 (2): 137–40. https://doi.org/10.1080/07491409.2016.1176788.

FDEP (Florida Department of Environmental Protection). 2018. *Volusia Blue Spring Basin Management Action Plan (BMAP)*. https://floridadep.gov/dear/water-quality-restoration/documents/volusia-blue-spring-basin-management-action-plan-bmap.

———. 2019. "FDEP Public Forum: Septic Tank Upgrade Incentives." Wednesday, January 30, 2019, 6:30PM." https://www.deltonafl.gov/deltona-tv/pages/fdep-forum-septic-tank-upgrades.

———. 2021a. "Manatees at Blue Spring State Park. Winter Home of Manatees." https://www.floridastateparks.org/index.php/parks-and-trails/blue-spring-state-park/manatees-blue-spring-state-park.

———. 2021b. "Septic Upgrade Incentive Program." https://floridadep.gov/springs/restoration-funding/content/septic-upgrade-incentive-program.

——— n.d. *Septic Upgrade Incentive Program*, accessed August 11, 2020. https://www.swfwmd.state.fl.us/sites/default/files/medias/documents/Septic Upgrade Incentive Program Information.pdf.

Feezor Engineering, Inc. 2020. *Incident Management Plan With Contingency Plan and Emergency Procedures*. https://dnr.mo.gov/env/swmp/facilities/docs/incidentmanagementplanrevision03302020.pdf.

Fischer, Frank. 2005. *Citizens, Experts, and the Environment: The Politics of Local Knowledge*. Durham: Duke University Press.

Fish & Wildlife Foundation of Florida. 2015. "Stormwater Treatment Area 5/6." Trail Sites. https://floridabirdingtrail.com/trail/trail-sections/south-section/stormwater-treatment-area-5-6/.

Flamm, Richard Owen, John Elliot Reynolds III, and Craig Harmak. 2013. "Improving Conservation of Florida Manatees (Trichechus Manatus Latirostris): Conceptualization and Contributions Toward a Regional Warm-Water Network Management Strategy for Sustainable Winter Habitat." *Environmental Management* 51: 154–66. https://doi.org/10.1007/s00267-012-9985-4.

Florio, John, and Ouisie Shapiro. 2016. "America's Dirtiest Boat Races." *The New Yorker*. July 29, 2016. https://www.newyorker.com/sports/sporting-scene/americas-dirtiest-boat-races.

Foote, Stephanie, and Elizabeth Mazzolini, eds. 2012. *Histories of the Dustheap: Waste, Material Cultures, Social Justice*. Cambridge: MIT Press.
Foucault, Michel. (1967) 1986. "Of Other Spaces," translated by Jay Miskowiec. *Diacritics* 16 (1): 22–27.
———. (1971) 1994. *The Order of Things: An Archeology of the Human Sciences*. New York: Vintage Books.
Frost, Nicola. 2016. "Anthropology and Festivals: Festival Ecologies." *Ethnos* 81(4): 569–83. https://doi.org/10.1080/00141844.2014.989875.
Gadamer, Hans-Georg. 1997. "Rhetoric, Hermeneutics, and Ideology-Critique," translated by G. B. Hess and R. E. Palmer. In *Rhetoric and Hermeneutics in Our Time: A Reader*, edited by Walter Jost and Micheal J. Hyde, 313–34. New Haven: Yale University Press.
Genocchio, Benjamin. 1996. "Discourse, Discontinuity, Difference: The Question of 'Other' Spaces." In *Postmodern Cities and Spaces*, edited by Sophie Watson and Katherine Gibson, 35–46. Oxford: Blackwell.
Gibbs, Lois M. (1982) 2011. *Love Canal and the Birth of the Environmental Health Movement*. Washington D.C.: Island Press.
Giddens, Anthony. 1990. *The Consequences of Modernity*. Stanford: Stanford University Press.
Gottlieb, Robert. 2005. *Forcing the Spring: The Transformation of the American Environmental Movement*. Washington D.C.: Island Press.
Gray, Jennifer. 2016. "Fish Kill in Florida: 'Heartbreaking Images' Seen for Miles." CNN, March 25, 2016. https://www.cnn.com/2016/03/25/us/florida-fish-kill.
Gross, Alan G. 1994. "The Roles of Rhetoric in the Public Understanding of Science." *Public Understanding of Science* 3 (1): 3–23.
Gustavson, Karl E., G. Allen Burton, Norman R. Francingues Jr., Danny D. Reible, Donna J. Vorhees, and John R. Wolfe. 2008. "Evaluating the Effectiveness of Contaminated-Sediment Dredging. *Environmental Science and Technology* 42 (14): 5042–47. https://doi.org/10.1021/es087185a
Harrington, Debra, Gary Maddox, and Richard Hicks. (2010). *Florida Springs Initiative Monitoring Network Report and Recognized Sources of Nitrate*. Florida Department of Environmental Protection. https://floridadep.gov/dear/water-quality-evaluation-tmdl/documents/florida-springs-initiative-monitoring-network-report.
Havens, Karl E., and Dale E. Gawlik. 2005. "Lake Okeechobee Conceptual Ecological Model." *Wetlands* 25 (4): 908–925. https://doi.org/10.1672/0277-5212(2005)025[0908:LOCEM]2.0.CO;2.
Hawkins, Gay. 2006. *The Ethics of Waste: How We Relate to Rubbish*. Lanham, MD: Rowman & Littlefield.
Healy, Stephen. 2004. "A 'Post-Foundational' Interpretation of Risk: Risk as Performance." *Journal of Risk Research* 7 (3): 277–96. https://doi.org/10.1080/1366987042000176235.
Heise, Ursula K. 2008. *Sense of Place and Sense of Planet: The Environmental Imagination of the Global*. Oxford: Oxford University Press.
Hetherington, Kevin. 1997. *The Badlands of Modernity: Heterotopia and Social Ordering*. London: Routledge.

———. 2004. "Secondhandedness: Consumption, Disposal, and Absent Presence." *Environment and Planning D: Society and Space* 22 (1): 157–73. https://doi.org/10.1068/d315t.
Hird, Myra J. 2013. "Waste, Landfills, and an Environmental Ethic of Vulnerability." *Ethics & the Environment* 18 (1): 105–24. https://doi.org/10.2979/ethicsenviro.18.1.105.
IEPA (Illinois Environmental Protection Agency). 2016. *Depue/New Jersey Zinc/Mobil Chemical Site Summary*. Ecological Risks. https://www2.illinois.gov/epa/topics/community-relations/sites/new-jersey-zinc/Pages/default.aspx.
———. 2019a. *NJZ/OU5 Review of FNC in Situ Caged Fingernail Clam Study Results*. IEPA Document Explorer, Superfund Technical, March 11, 2019. https://external.epa.illinois.gov/DocumentExplorer/Documents/Index/170000002878.
———. 2019b. *NJZ/OU5, Evaluation of Responses to Comments/in-Situ Toxicity Test*. IEPA Document Explorer, Superfund Technical, September 16, 2019. https://external.epa.illinois.gov/DocumentExplorer/Documents/Index/170000002878.
———. 2019c. *Letter to Nancy Loeb*. IEPA Document Explorer, Superfund Technical, July 16, 2019. https://external.epa.illinois.gov/DocumentExplorer/Documents/Index/170000002878.
Johnson, Peter. 2006. "Unravelling Foucault's 'Different Spaces.'" *History of the Human Sciences* 19 (4): 75–90. https://doi.org/10.1177/0952695106069669.
Kasperson, Jeanne X. and Roger E. Kasperson. 2005. *The Social Contours of Risk: Publics, Risk Communication and the Social Amplification of Risk, Volume 1*. London: Earthscan.
Kernc, Christopher. 2011. "Black Diamonds: The Rise and Fall of the Northern Illinois Coal Industry." *Historia* 20: 71–83.
Keyser, Jason. 2012. "Amid Drought, Volunteer Saves Illinois Boat Race." *Journal Star*, July 26, 2012. https://www.pjstar.com/article/20120726/NEWS/307269866.
Kirk, Andrew. 2012. "Rereading the Nature of Atomic Doom Towns." *Environmental History* 17: 635–647. https://doi.org/10.1093/envhis/ems049.
Knezevic, Irena. 2009. "Hunting and Environmentalism: Conflict or Misperceptions." *Human Dimensions of Wildlife* 14 (1): 12–20. https://doi.org/10.1080/10871200802562372.
Krupar, Jason. 2007. "Burying Atomic History: The Mound Builders of Fernald and Weldon Spring." *The Public Historian* 29 (1): 31–58. https://doi.org/10.1525/tph.2007.29.1.31.
Krupar, Jason, and Stephen DePoe. 2007. "Cold War Triumphant: The Rhetorical Uses of History, Memory, and Heritage Preservation Within the Department of Energy's Nuclear Weapons Complex." In *Nuclear Legacies: Communication, Controversy, and the U.S. Nuclear Weapons Complex*, edited by Bryan C. Taylor, William J. Kinsella, Stephen P. DePoe, and Maribeth S. Metzler, 135–66. Lanham, MD: Lexington Books.
Laclau, Ernesto, and Chantal Mouffe. 2001. *Hegemony and Socialist Strategy: Towards a Radical Democratic Politics*. second ed. London: Verso.
Laist, David W., and John E. Reynolds III. 2005. "Florida Manatees, Warm-Water Refuges, and an Uncertain Future." *Coastal Management* 33 (3): 279–295. https://doi.org/10.1080/08920750590952018.

Lapointe, Brian E., Laura W. Herren, David D. Debortoli, and Margaret A. Vogel. 2015. "Evidence of Sewage-Driven Eutrophication and Harmful Algae Blooms in Florida's Indian River Lagoon." *Harmful Algae* 43: 82–102. https://doi.org/10.1016/j.hal.2015.01.004.

Lash, Scott. 1994. "Reflexivity and Its Doubles: Structure, Aesthetics, Community." In *Reflexive Modernization: Politics, Tradition and Aesthetics in the Modern Social Order*, edited by Ulrich Beck, Anthony Giddens, and Scott Lash, 110–173. Stanford: Stanford University Press.

Latour, Bruno. 2011. "Love Your Monsters: Why We Must Care For Our Technologies As We Do Our Children." *Breakthrough Journal* 2: 19–26.

Law, John, and Vicky Singleton. 2005. "Object Lessons." *Organization* 12 (3): 331–355. https://doi.org/10.1177/1350508405051270.

Lee, Ming T., and John B. Stall. 1976. *Sediment Conditions in Backwater Lakes Along the Illinois River—Phase 2*. Contract Report 176. State of Illinois. Department of Registration and Education. https://www.isws.illinois.edu/pubdoc/CR/ISWSCR-176-B.pdf.

Massey, Doreen. (2005) 2015. *For Space*. Los Angeles: Sage.

Mentz, Steve. 2012. "After Sustainability." *PMLA* 127 (3): 586–592.

Milstein, Tema. 2008. When Whales 'Speak for Themselves:' Communication as a Mediating Force in Wildlife Tourism. *Environmental Communication* 2 (2): 173–192. https://doi.org/10.1080/17524030802141745.

Mol, Annemarie. 2002. *The Body Multiple: Ontology in Medical Practice*. Durham: Duke University Press.

Mol, Annemarie, and John Law. 1994. "Regions, Networks and Fluids: Anaemia and Social Topology." *Social Studies of Science* 24 (4): 641–71. https://doi.org/10.1177/030631279402400402.

Moore, Sarah. 2011. "Village of Depue Receives Response From IEPA About Lake Cleanup." September 21, 2011. *Northwestern University Global Health Portal*. https://blog.globalhealthportal.northwestern.edu/2011/09/village-of-depue-receives-response-from-iepa-about-lake-cleanup/.

Moscardo, Gianna, Barbara Woods, and Rebecca Saltzer. 2004. "The Role of Interpretation in Wildlife Tourism." In *Wildlife Tourism: Impacts Management Planning*, edited by Karen Higginbottom, 231–251. Queensland: Common Ground Publishing.

Mosley, Stephen. 2014. "Environmental History of Air Pollution and Protection." In *The Basic Environmental History: Environmental History Vol. 4*, edited by Mauro Agnoletti and Simone Neri Serneri, 143–71. Heidelberg: Springer Cham.

National Research Council. 2003. *Bioavailability of Contaminants in Soils and Sediments: Processes, Tools, and Applications*. Washington D.C.: The National Academies Press. https://doi.org/10.17226/10523.

The Nature Conservancy. 2021a. "Stories in Illinois. Water Management at Emiquon." https://www.nature.org/en-us/about-us/where-we-work/united-states/illinois/stories-in-illinois/water-management-faq/.

The Nature Conservancy. 2021b. "Places We Protect. Emiquon, Illinois." https://www.nature.org/en-us/get-involved/how-to-help/places-we-protect/emiquon/.

NCICG (North Central Illinois Council of Governments). 2014. *Village of Depue Comprehensive Plan 2014*. https://www.ncicg.org/planning-services/comprehensive-plans/depue-comprehensive-plan/.

Newell, Stephanie. 2015. "Dirty Familiars: Colonial Encounters in African Cities." In *Global Garbage: Urban Imaginaries of Excess, Waste and Abandonment in the Contemporary City*, edited by Christoph Lindner and Miriam Meissner, 35–51. New York: Routledge.

Novotny, Patrick. 2000. *Where We Live, Work and Play: The Environmental Justice Movement and the Struggle for a New Environmentalism*. Westport: Praeger.

Nye, David E. 1994. *American Technological Sublime*. Cambridge: MIT Press.

OLM (Office of Legacy Management). n.d. "Sites: Office of Legacy Management." Accessed September 20, 2021. https://www.energy.gov/lm/sites

———. 2005. *Long-Term Surveillance and Maintenance Plan for the U. S. Department of Energy Weldon Spring, Missouri, Site*. https://www.lm.doe.gov/Weldon/final_weldon_ltsp.pdf.

———. 2015a. "Weldon Spring Site Interpretive Online Tour, Disposal Cell Construction. Cell Design." http://www.lm.doe.gov/Weldon/Interpretive_Center/Online_Tour/Disposal_Cell_Construction.pdf.

———. 2015b. "Weldon Spring Site Interpretive Online Tour, Disposal Cell Mural. Cover." https://www.lm.doe.gov/Weldon/Interpretive_Center/Online_Tour/Disposal_Cell_Mural.pdf.

———. 2015c. "Weldon Spring Site Interpretive Online Tour, Radiation Fundamentals. How Does One Protect Oneself from Radiation?" https://www.lm.doe.gov/Weldon/Interpretive_Center/Online_Tour/Radiation_Fundamentals.pdf.

———. 2015d. "Weldon Spring Site Interpretive Online Tour, Missouri Prairies. What is Howell Prairie?" https://www.lm.doe.gov/Weldon/Interpretive_Center/Online_Tour/Missouri_Prairies.pdf.

Oravec, Christine. 1982. "The Evolutionary Sublime and the Essay of Natural History." *Communication Monographs* 49 (4): 216–227. https://doi.org/10.1080/03637758209376086.

Osborne, Susan, and Jessica Blunden. 2013. "2012—Third Warmest La Nina on Record." NOAA Climate.gov. January 15, 2013. https://www.climate.gov/news-features/featured-images/2012%E2%80%94third-warmest-la-ni%C3%B1a-year-record.

Peeples, Jennifer. 2011a. "Toxic Sublime: Imaging Contaminated Landscapes." *Environmental Communication* 5 (4): 373–392. https://doi.org/10.1080/17524032.2011.616516.

———. 2011b. "Downwind: Articulation and Appropriation of Social Movement Discourse." *Southern Communication Journal* 76 (3): 248–263. https://doi.org/10.1080/1041794x.2010.500516.

Pezzullo, Phaedra C. 2007. *Toxic Tourism: Rhetorics of Pollution, Travel, and Environmental Justice*. Tuscaloosa: University of Alabama Press.

———. 2012. "What Gets Buried in a Small Town: Toxic E-waste and Democratic Frictions in the Crossroads of the United States." In *Histories of the Dustheap: Waste, Material Cultures, Social Justice*, edited by Stephanie Foote and Elizabeth Mazzolini, 119–145. Cambridge: MIT Press.

Picard, David. 2016. "The Festive Frame: Festivals as Mediators for Social Change." *Ethnos* 81 (4): 600–616. https://doi.org/10.1080/00141844.2014.9898 69.

Pløger, John. 2010. "Presence-Experiences—the Eventalisation of Urban Space." *Environment and Planning D: Society and Space* 28 (5): 848–866. https://doi.org/10.1068/d1009.

Prelli, Lawrence J. 2006. "Rhetorics of Display: An Introduction." In *Rhetorics of Display*, edited by Lawrence J. Prelli, 1–38. Columbia: University of South Carolina Press.

Quinn, Bernadette, and Linda Wilks. 2017. "Festival Heterotopias: Spatial and Temporal Transformations in Two Small-Scale Settlements." *Journal of Rural Studies* 53: 35–44. https://doi.org/10.1016/j.jrurstud.2017.05.006.

Reese, Erik. 2007. "Reclaiming a Toxic Legacy Through Art and Science." *Orion Magazine* 26 (6). https://orionmagazine.org/article/reclaiming-a-toxic-legacy-through-art-and-science/.

Reeves, Joshua. 2013. "Suspended Identification: Atopos and the Work of Public Memory." *Philosophy & Rhetoric* 46 (3): 306–327.

Rekret, Paul. 2016. "A Critique of New Materialism: Ethics and Ontology. *Subjectivity* 9 (3): 225–245. https://doi.org/10.1057/s41286-016-0001-y.

Rickert, Thomas. 2013. *Ambient Rhetoric: The Attunements of Rhetorical Being*. Pittsburgh: University of Pittsburgh Press.

Rogers, Heather. 2006. *Gone Tomorrow: The Hidden Life of Garbage*. New York: The New Press.

Rose, Arthur W. 2013. *An Evaluation of Passive Treatment Systems Receiving Oxic Net Acidic Mine Drainage*. Datashed. https://datashed.org/sites/default/files/art_rose_passive_system_evaluations_2013.pdf.

Royte. Elizabeth. 2005. *Garbage Land: On the Secret Trail of Trash*. New York: Back Bay Books/Little, Brown and Company.

Runge, Michael C., Carol A. Sanders-Reed, Catherine A. Langtimm, Jeffrey A. Hostetler, Julien Martin, Charles J. Deutsch, Leslie I. Ward-Geiger, and Gary L. Mahon. 2017. *Status and Threats Analysis for the Florida Manatee (Trichechus Manatus Latirostris), 2016*. U.S. Geological Survey Scientific Investigation Report 2017-5030, 40 p. https://doi.org/10.3133/sir20175030.

Saindon, Brent Allen. 2010. "A Doubled Heterotopia: Shifting Spatial and Visual Symbolism in the Jewish Museum Berlin's Development." *Quarterly Journal of Speech* 98 (1): 24–48. https://doi.org/10.1080/00335630.2011.638657.

Schade-Poole, Kristin, and Gregory Möller. 2016. "Impact and Mitigation of Nutrient Pollution and Overland Water Flow Change on the Florida Everglades, USA." *Sustainability* 8 (9) (940): 1–20. https://doi.org/10.3390/su8090940.

Schlosberg, David. 2013. "Theorising Environmental Justice: The Expanding Sphere of Discourse." *Environmental Politics* 22 (1): 37–55. https://doi.org/10.1080/09644016.2013.755387.

Schneider, Daniel. 2012. "Purification or Profit: Milwaukee and the Contradictions of Sludge." In *Histories of the Dustheap: Waste, Material Cultures, Social Justice*, edited by Stephanie Foote and Elizabeth Mazzolini, 171–197. Cambridge: MIT Press.

Shute, Kim. 2013. "Village Wants to Separate Lake Depue From River." *News-Tribune*, May 23, 2013.

———. 2016. "Depue: DNR Says OK to Temporary Dam." *NewsTribune*, May 25, 2016. https://www.shawlocal.com/articles/tn/2016/05/25/ef92646f06975e73bc81eea4221c7ef7.

Senda-Cook, Samantha. 2013. "Materializing Tensions: How Maps and Trails Mediate Nature." *Environmental Communication* 7 (3): 355–71. https://doi.org/10.1080/17524032.2013.792854.

SFWMD (South Florida Water Management District). n.d. "Progress Continues on the Everglades Agricultural Area Reservoir Project." Accessed, May 20, 2020. https://www.sfwmd.gov/our-work/cerp-project-planning/eaa-reservoir.

———. n.d. "Harmful Nutrients in the Everglades Now Reduced by 90%." Improving Everglades Water Quality. Accessed, May 20, 2020. https://www.sfwmd.gov/our-work/wq-stas.

———. n.d. "Water Quality Improvement." Stormwater Treatment Areas (STAs). Accessed, May 20, 2020. https://www.sfwmd.gov/our-work/wq-stas.

———. n.d. "Storm Water Treatment Area 5/6 (STA 5/6)." Accessed, May 20, 2020. https://www.sfwmd.gov/recreation-site/stormwater-treatment-area-56-sta-56.

———. n.d. "Lakeside Ranch Stormwater Treatment Area, FYI." Accessed, June 2, 2021. https://www.sfwmd.gov/sites/default/files/documents/lowpp_tcns_s191_mm51-55_sheet.pdf.

———. 2018. *Central Everglades Planning Project Post Authorization Change Report: Feasibility Study and Draft Environmental Impact Statement*. https://www.sfwmd.gov/sites/default/files/documents/cepp_pacr_main_report.pdf.

Simon, Stephanie. 2002. "From Atomic Dump to Tourist Draw." *Los Angeles Times*, August 13, 2002. https://www.latimes.com/archives/la-xpm-2002-aug-13-na-toxic13-story.html.

SJRWMD (Saint John's River Water Management District). 2020. "The Indian River Lagoon." https://www.sjrwmd.com/waterways/indian-river-lagoon/.

Sohn, Heidi. 2008. "Heterotopia: Anamnesis of a Medical Term." In *Heterotopia and the City: Public Space in a Postcivil Society*, edited by Michiel Dehaene and Lieven De Cauter, 41–50. New York: Routledge.

Smith, Wes. 1996. "Heaps of History Captured in Piles of Coal Mine Rock." *Chicago Tribune*, December 2, 1996. https://www.chicagotribune.com/news/ct-xpm-1996-12-02-9612020079-story.html.

Spirn, Anne Whiston. 1996. "Constructing Nature: The Legacy of Frederick Law Olmsted." In *Uncommon Ground: Rethinking the Human Place in Nature*, edited by William Cronon, 91–113. New York: W. W. Norton & Company.

Stern, Marcus, and Meryl Kornfeld. 2020. "Why Florida's Toxic Algae Crisis is Worse Than People Realize." *Tampa Bay Times*, June 10, 2020. https://www.tampabay.com/news/environment/2020/06/08/why-floridas-toxic-algae-crisis-is-worse-than-people-realize/.

Stone, Philip R. 2013. "Dark Tourism, Heterotopias and Post-Apocalyptic Places: The Case of Chernobyl." In *Dark Tourism and Place Identity*, edited by Leanne White and Elspeth Frew, 79–94. Melbourne: Routledge.

Stormer, Nathan and Bridie McGreavy 2017. "Thinking Ecologically About Rhetoric's Ontology: Capacity, Vulnerability, and Resilience." *Philosophy & Rhetoric* 50 (1): 1–25. https://doi.org/10.5325/philrhet.50.1.0001.

Swain, Hilary M., Patrick J. Bohlen, Kenneth L. Campbell, Laurent O. Lollis, and Alan D. Steinman. 2007. "Integrated Ecological and Economic Analysis of Ranch Management Systems: An Example From South Central Florida." *Rangeland Ecology & Management* 60 (1): 1–11. https://doi.org/10.2111/05-071R1.1.

Swanson, Jess. 2016. "Five Powerful Photos Show Massive Fish Kill in Florida." *Broward Palm Beach New Times*, March 23, 2016. https://www.browardpalmbeach.com/news/five-powerful-photos-show-massive-fish-kill-in-florida-7670175.

Taylor, Bryan C. 2010. "Radioactive History: Rhetoric, Memory, and Place in the Post-Cold War Nuclear Museum." *Places of Public Memory: The Rhetoric of Museums and Memorials*, edited by Greg Dickinson, Carole Blair, and Brian L. Ott, 57–86. Tuscaloosa: University of Alabama Press.

TECO (Tampa Electric Company). 2016. "Big Bend Power Station. Clean Air Technology: White Plumes From Big Bend Power Station Stacks." Interpretive Panel, Manatee Viewing Center Environmental Education Center, Apollo Beach FL.

———. 2021. "Tampa Electric's Manatee Viewing Center." https://www.tampaelectric.com/company/mvc/.

Thornes, John E. 2008. "A Rough Guide to Environmental Art." *Annual Review of Environment and Resources* 33: 391–411. https://doi.org/10.1146/annurev.environ.31.042605.134920.

Topinka, Robert J. 2010. "Foucault, Borges, Heterotopia: Producing Knowledge in Other Spaces." *Foucault Studies* 9: 54–70. https://doi.org/10.22439/fs.v0i9.3059.

Tripp, Katie. 2016. "Save the Manatee Club's Official Comments to the U.S. Fish and Wildlife Service." Save the Manatee Club. Opposition to Manatee Downlisting. April 7, 2016. https://www.savethemanatee.org/wp-content/uploads/2016/05/fws_Downlisting_Comment_KT_4-16.pdf.

Tulloch, John. 2008. "Culture and Risk." In *Social Theories of Risk and Uncertainty: An Introduction*, edited by Jens O. Zinn, 138–167. Malden: Blackwell Publishing.

USACE (United States Army Corps of Engineers). 2007. *Illinois River Basin Restoration Comprehensive Plan With Integrated Environmental Assessment.* www.mvr.usace.army.mil/Portals/48/docs/Environmental/ILRBR/CompPlan/Main Report and Appendices - CD.pdf.

U.S. Census Bureau. 2019. *Quick Facts, Deltona City, Florida.* https://www.census.gov/quickfacts/deltonacityflorida.

———. 2019. *Quick Facts, Volusia County, Florida.* https://www.census.gov/quickfacts/volusiacountyflorida.

USEPA (United States Environmental Protection Agency). 2018. "West Lake Landfill Going Forward. Isolation Barrier System." https://archive.epa.gov/epa/mo/west-lake-landfill-going-forward.html.

———. 2021a. "Westlake Landfill Bridgeton, MO Cleanup Activities. Background." https://cumulis.epa.gov/supercpad/SiteProfiles/index.cfm?fuseaction=second.Cleanup&id=0701039#bkground.

———. 2021b. "National Priorities List (NPL) Sites—By State." https://www.epa.gov/superfund/national-priorities-list-npl-sites-state.

USFWS (U.S. Fish & Wildlife Service). 2017. "Endangered and Threatened Wildlife and Plants: Reclassification of the West Indian Manatee From Endangered to Threatened." *Federal Register* 82 (64): 16668–16704. Accessed from: https://www.federalregister.gov/documents/2017/04/05/2017-06657/endangered-and-threatened-wildlife-and-plants-reclassification-of-the-west-indian-manatee-from.
van den Berg, Hubert F. 2006. "Towards a 'Reconciliation of Man and Nature'. Nature and Ecology in the Aesthetic Avant-Garde of the Twentieth Century." In *Neo-Avant-Garde (Avant Garde Critical Studies Vol. 20)*, edited by David Hopkins, 371–387. Amsterdam: Rodopi.
Vos, Robert O. 2007. "Perspective. Defining Sustainability: A Conceptual Orientation. *Journal of Chemical Technology and Biotechnology* 82 (4): 334–339. https://doi.org/10.1002/jctb.1675.
Walsh, Lynda, Nathaniel A. Rivers, Jenny Rice, Laurie E. Gries, Jennifer L. Bay, Thomas Rickert, and Carolyn R. Miller. 2017. "Forum: Bruno Latour on Rhetoric." *Rhetoric Society Quarterly* 47 (5): 403–462. https://doi.org/10.1080/02773945.2017.1369822.
Washick, Bonnie, Elizabeth Wingrove, Kathy E. Ferguson, and Jane Bennett. 2015. "Politics That Matter: Thinking About Power and Justice with the New Materialists." *Contemporary Political Theory* 14 (1): 63–89. https://doi.org/10.1057/cpt.2014.19.
Webber, Tammy. 2011. "Depue Impatient for Answers on Pollution Cleanup." *The State Journal Register*. January 16, 2011. https://www.sj-r.com/article/20110116/NEWS/301169967.
Weller, Sally. 2013. "Consuming the City: Public Fashion Festivals and the Participatory Economies of Urban Spaces in Melbourne, Australia." *Urban Studies* 50 (14): 2853–2868. https://doi.org/10.1177/0042098013482500.
Wells, Justine, Bridie McGreavy, Samantha Senda-Cook, and George F. McHendry. Jr. 2018. "Introduction: Rhetoric's Ecologies." In *Tracing Rhetoric and Material Life: Ecological Approaches*, edited by Bridie McGreavy, Justine Wells, George. F. McHendry. Jr., and Samantha Senda-Cook, 1–36. Cham: Palgrave Macmillan.
Wesselman, Daan. 2013. "The High Line, 'The Balloon,' and Heterotopia." *Space and Culture* 16 (1): 16–27. https://doi.org/10.1177/1206331212451669.
Williams, Raymond. 1980. *Problems in Materialism and Culture: Selected Essays*. London: Verso.
WCED (World Commission on Environment and Development). 1987. *Our Common Future*. New York: United Nations. https://digitallibrary.un.org/record/139811?ln=en.
Zinn, Jens O. 2008b. "A Comparison of Sociological Theorizing on Risk and Uncertainty." In *Social Theories of Risk and Uncertainty: An Introduction*, edited by Jens O. Zinn, 168–210. Malden: Blackwell Publishing.

Index

acid mine drainage (AMD), 8–9, 99–103, 106, 113, 150
Ackerman, John, 40
agency, 9, 24, 26, 31, 55, 59, 105, 128; heterogeneous agents, 56; human agency, tendency to privilege, 30, 58; material assemblages, granting agency to, 24; new materialism and, 25–26, 40, 120; of nonhuman matter, 3, 23, 82, 106; public agency in festival spaces, 126–28, 133–35, 138, 146; rhetorical agency, 44, 50; sedimentation and human agency, 131–33; Superfund sites, limited agency at, 121
Ahsapa Water Control Structure, 147
air pollution, 18–19, 22
Alaimo, Stacy, 59
Amigoni, Joe, 34
Anderson, Ben, 127
apocalyptic rhetoric, 27–28, 73
ART & AMD Project, 99–103, 106, 113, 150
Ashe, Arthur, 44
Atomic Energy Commission, 103
atopos, notion of, 45, 46, 50–51
Audubon Florida organization, 78

Banklick Creek, 100
Barad, Karen, 19, 26
Barker, Jacob, 115
Barnett, Joshua Trey, 97
baseline ecological risk assessment (BERA), 136
Basin Management Action Plans (BMAPs), 68
Beck, Ulrich, 3, 5–6, 20, 81, 122–23
Bennett, Jane, 13–14, 25, 26–27, 30–31, 41, 65
Benson, Melinda Harm, 56, 59–60, 67, 152
Bergthaller, Hannes, 26, 30, 73
Big Bend Power Plant, 61, 62, 68; decontamination of, 66–67; as a heterotopic space, 53–54, 82–83; posthuman hermeneutics and, 55–56, 60; sustainability, as a site of, 64, 82, 128
bioavailability, 136–38
Bitzer, Lloyd F., 48–49
Blair, Carole, 42, 44–45
blue-green algae, 70–73, 74–77, 85–86
Blue Spring State Park, 61, 63, 67–68, 69
Borges, Jorge Luis, 38–39

167

Braidotti, Rosi, 28, 32, 55–56
Bridgeton Landfill, 114, 116
Broward Palm Beach New Times (periodical), 72
Brown, Kate, 33
Bryant, Eric, 130, 132, 133–35
Buell, Lawrence, 5–6, 132
Burtynsky, Edward, 92, 98, 101, 106

Callahan, Richard J., Jr., 15
Caloosahatchee River, 72, 76
Carr, John, 53–55, 62, 82
Carson, Rachel, 27
cattails, 79–80
cattle industry, 75, 76
Chernobyl, site of, 94
Civil Rights Memorial, 44–45
Clean Air Act (CAA), 18–19, 20, 21
"cleanliness as next to godliness" slogan, 15–16
The Cleanliness Institute, 15–16
Clean Water Act (CWA), 20, 21
climate change, 10, 29, 59, 60, 62, 67, 152
Coal Miner's Park, 35, 51
commemoration, sites of, 42, 44, 45, 126
community engagement, energy of, 100, 102–3
Comp, T. Allen, 101
Comprehensive Emergency Response, Compensation, and Liability Act (CERCLA), 21, 103. *See also* Superfund
Confederate statues of Monument Avenue, 42, 44
Coole, Diana, 24
corporate greenwashing, 62, 82
Cortez, Marisol, 12
Craig, Robin Kundis, 56, 59–60, 67, 152
Critical Mass bike rides, 46–47
Crystal River, 61
Cuyahoga River, 20

De Certeau, Michel, 47
decontamination, 60, 66

Deltona, city of, 68, 69, 72–73, 74, 85–86
DeLuca, Kevin Michael, 43–44, 57, 96, 124, 139
Demo, Teresa, 57, 96
Denora, town of, 18, 22
Department of Energy (DOE), 89, 104, 112–13
Dickinson, Greg, 42
dirty matter, 3, 17, 22, 24, 68, 72, 96, 105; community engagement with, 102–3; elimination and removal of, 55, 106; as ever-present, 23, 27, 114, 151; exigence as linked with, 7, 48; familiar environmental habits and, 10, 82, 91; new modes of perception for, 4, 28, 36, 98; novel configurations of, 2, 7; OSTDS association with, 71; perceived absence of, 6, 35, 116, 149; recreation sites, transformation into, 1, 142; recycling not always available for, 25
dissonance, 23, 32, 36, 41, 50, 94, 101; at Big Bend Power Plant, 54; dirty configurations as constituting, 55, 150; from dirty world deliberations, 140; heterotopic dissonance, 46, 49; of initial encounter with the sublime, 106; Land Art and spaces of modern dissonance, 98; of massive technical object encounters, 95; multiplicity and dissonance, perceiving space through, 56; perceptual and responsive tendency to, 45; pervasiveness of waste, dissonance generated by, 5, 48, 50; postmodern risk, dissonance inherent to, 113; technocentrism as a mediating tool for, 58–59; of twenty-first century spaces, 2, 7; at Weldon Spring Site, 110, 117
Douglas, Mary, 15, 16
downwind/downstream, tropes of, 8, 74, 86, 124, 150
Drainage Acts of Florida (1916), 76

dredging, 9, 10, 132, 135, 137–38, 140–48
Ducks Unlimited hunting association, 139

E² and D.I.R.T. Studio report, 143–45, 147–48
earth, space images of, 32, 58, 86
eco-tourism, 94, 143
Edbauer, Jenny, 36
electricity blackout of 2003, 65
emergent causality, 26
Emiquon Preserve, 146–48
Endres, Danielle, 46
environmental justice, 1, 32, 43, 86, 94, 117; class relations and, 12, 115–16; environmentalism and, 135, 139–40, 142; public agency and, 122, 138; public participation in, 114, 151; rearticulation of the "environment" as central to, 124–25
environmental photography, 92, 96–97
Environmental Protection Agency (EPA), 116–17, 120, 121, 136–37
eutrophication, 19–20, 75
Everglades, 76, 79, 80, 151
Everglades Agricultural Area (EAA), 77
Everglades Agricultural Area Storage Reservoir Project, 77–78

festival spaces. *See* Lake Depue
fingernail clam (FNC) study, 136–37
first order springs, 67
Fischer, Frank, 122, 123
Fish & Wildlife Foundation of Florida, 79
Flamm, Richard Owen, 66
Florida Department of Environmental Protection (FDEP), 68, 69
Floridan Aquifer, 67
Florida Springs and Aquifer Act, 68
flow equalization basins (FEB), 78
Foucault, Michel, 2, 14, 29, 37–39, 94
Frost, Nicola, 24, 126

Gadamer, Hans-Georg, 45
Genocchio, Benjamin, 38
Gibbs, Lois, 20–21, 44, 124
Giddens, Anthony, 145–46
The Gleaners and I (film), 17
Gottlieb, Robert, 20
Great Lakes, 19–20
Great Salt Lake, 91, 98–99
Gross, Alan G., 105

Hamburg Trail, 104
happenstance, 9, 74, 82, 102, 127, 148; acceptance of agency and, 106; in festival spaces, 9, 132; surprise, element of, 47
Harmak, Craig, 66
Hawkins, Gay, 12–13, 17
Heidegger, Martin, 59
heterotopia, 6, 7, 31, 40, 56, 62, 94, 128, 142; alternative influence of waste, 47–52; AMD & ART Project as a heterotopic space, 100, 103; Big Bend as a heterotopic space, 53–55, 82–83; Burtynsky, heterotopic quality of images, 98; as creative space, 37–41; describing and defining, 1, 2, 5, 14, 36–37; ethics and, 3, 9; of festival spaces, 2, 130; heterotopic dissonance, 46, 49, 98; incongruity of, 4, 65, 79, 121; as juxtaposition of discordant sites, 29, 99; MVC, heterotopic configuration of, 64, 66; New Topographic movement and, 92; posthuman heterotopia, 32, 60, 81, 90, 106, 116; public participation in a dirty world, 10, 150; Weldon Spring Site as a heterotopic space, 8, 90–91
Hetherington, Kevin, 14, 16
Hooker Chemical, 20
Howell Prairie, 104, 111, 112
hydroelectric power, 20

Illinois Department of Natural Resources (IDNR), 135
Illinois Environmental Protection Agency (IEPA), 120, 136–37, 140, 145

Illinois River, 10, 33, 143, 146; floodplain of, 119, 128, 135; Lake Depue as draining into, 131–32; redirection of water into, 144–45; restoration practices along, 142; stressors impacting, 147–48; views over the river, 35, 51
Indian River Lagoon (IRL), 72, 74–75, 84, 85
intra-action, 26, 125, 127, 136

Johnson, Peter, 38
jumbos (slag piles), 34
JustMomsStL advocacy group, 116–17

Kasperson, Jeanne X. and Roger E., 104
Katy Trail, 90, 104
Kissimmee watershed, 75
Knezevic, Irena, 139

Laclau, Ernesto, 43
Lake Depue: as contaminated, 120, 125–26, 133; festival space at, 128–30, 133–35, 140, 146; Illinois River as source for, 144–45; speedboat races on, 2, 9, 119, 130, 131–32, 133–35, 138–42, 146; toxicity of, 136–37, 141, 143, 146; Village of Depue Comprehensive Plan for, 134–35; wetland, plans for conversion into, 144, 147–48, 151
Lake Erie, 20, 22
Lake Michigan, 144–45
Lake Okeechobee, 70, 72, 75–76, 77–81, 85, 144
Lakeside Ranch Storm Treatment, 81
Land Art, 91, 98, 101, 106
Lash, Scott, 123, 124
Latour, Bruno, 29, 31, 62
LeFevre, Gregory, 40
legacy pollution, 76
locomotives as objects of awe, 95, 96
Lofton, Kathryn, 15
Love Canal neighborhood, 20–21
Love, William T., 20–21

Mackle Brothers development, 69
Mallinckrodt Chemical co., 114–15
manatees, 70, 85; at Blue Spring State Park, 67–68; boardwalk, viewing manatees from, 63–64, 65, 68; in Deltona waters, 72–73, 86; Florida manatees, 53, 60, 75, 84; Save the Manatees (STM), 83–84
Manatee Viewing Center (MVC), 61–66, 68, 74, 79, 115, 117, 151
Massey, Doreen, 47, 127
McGreavy, Bridie, 4
Mentz, Steve, 56
Michel, Neil, 45
Miller, Carolyn R., 49–51, 56, 73
Milstein, Tema, 53–55, 62, 82
Missouri Department of Natural Resources (MDNR), 116
Missouri River, 90, 103, 104
Monongahela River, 18
Moscardo, Gianna, 63–64
Mouffe, Chantal, 43

natura naturans principle, 92
The Nature Conservancy, 146–48
new materialism, 2, 7, 13, 24, 73, 105; agency, perspective on, 25, 31, 40, 120, 138; change, new understanding of, 29–30; Ranciere, materialist reading of, 26–27; rhetorical theory as informed by, 43, 82; snake intruder example of revision, 49–51; waste, new materialist orientation to, 106
New Topographics, 91, 92
Niagara Falls, 20, 57
La Niña weather pattern, 131
nitrates, 66, 68–73, 75–76, 79, 84–86
Nye, David E., 95

Office of Legacy Management (OLM); prairie, describing, 111–12; the public, efforts to engage, 112–13; at Weldon Spring Site, 11, 89–90, 104, 108, 115
Of Other Spaces (Foucault), 37

operational units (OU), 120, 129
Oravec, Christine, 106
The Order of Things (Foucault), 38–39
OSTDS. *See* residential septic systems
Ott, Brian L., 42
Outstanding Florida Springs (OFS), 68

PCB chemicals, 142, 145
Peeples, Jennifer, 92, 93, 97–98, 101, 124
Pezzullo, Phaedra C., 93, 94, 96, 123
phosphorous, 20, 22, 75–81
pink shale, 33–34, 51
pragmatic acceptance, position of, 146
Prelli, Lawrence J., 42, 44
Primary Focus Areas (PFAs), 68, 69–70
principle responsible parties (PRP), 120, 121–22, 132, 133, 136–37, 140

Rancière, Jacques, 26–27
recirculation, 12–13
Reese, Erik, 101–2
Reeves, Joshua, 45–46
reflexive modernization, 3, 123
Rekret, Paul, 30
relationality, 2, 9, 14, 65, 74, 85; in a dirty space, 7, 8, 84, 118; in ecology of waste, 150–51; ethics and, 30, 83; heterotopic quality of, 4, 39, 73; lake, relationality in dirtiness of, 10, 148; posthuman view of, 55, 86
remediation, 1, 18, 23, 60, 85, 100, 144; of human waste via septic systems, 69, 70; at Lake Depue, 121–22, 129, 135, 145; revising practices of, 35–36, 106; of Spring Valley slag heaps, 34–35; of Weldon Spring Site, 8, 11, 89–90, 116
reorientation, 46, 51, 91, 117, 125; ethic of care, reorientation as enabling, 117, 142; heterotopia, perceptual reorientations due to, 40, 121; to presence of waste, 3, 66–67, 112; Weldon Spring as inspiring movement for, 111, 150

residential septic systems (OSTDS), 69–70, 71, 73, 77, 85
Resource Conservation and Recovery Act (RCRA), 21–22
Reynolds, John Elliot, III, 66
"The Rhetorical Situation" (Bitzer), 48
rhetoric and actor-network theory, 56
rhetoric of accommodation, 105
Rickert, Thomas, 48, 50, 59
risk and relational ethics, 114–18
risk society, 3, 5–6, 81, 104, 123
Rose, Arthur W., 102

"Safe Side of the Fence" (film), 115
Saltzer, Rebecca, 63–64
Sanitary and Ship Canal, 145
Schlosberg, Daniel, 124
Schneider, Daniel, 17–18
Seales, Chad E., 15
Senda-Cook, Samantha, 46
septic systems, 69–74, 75, 77, 78, 86
Septic Upgrade Incentive Program, 70
Smithson, Robert, 91, 98–99
soap and suds, 15–16, 19
social avoidance of waste, 17
South Florida Water Management District (SFWMD), 77, 78–79
speed boat races. *See* Lake Depue
"Spiral Jetty" (art piece), 91, 98–99
springsheds, 67, 68
Spring Valley, Illinois, 33–35, 51
St. Charles County, 103, 104, 111
St. John's River Watershed, 85–86
St. Lucie River, 72, 76
Stone, Philip R., 94
Stormer, Nathan, 4
Stormwater Treatment Areas (STAs), 77–81, 82, 85, 86, 144
subsurface smoldering event (SSE), 116
Sullivan, Robert, 25
Superfund, 21, 132, 141, 142; Superfund Act, 23; Superfund Redevelopment Initiative, 143; Superfund sites, 2, 103, 114, 119–20, 121, 129, 144

sustainability, 65, 66, 87, 92; Big Bend, as enacted in, 55, 82, 128; describing and defining, 8, 56; dissonance and, 59; MVC discourse on, 62–64, 74; *Our Common Future* as founding document of, 58

Tampa Bay, 53
Tampa Electric Company (TECO), 53, 61, 64, 128
technocentrism, 58
the technological sublime, 95, 106
technology forcing, process of, 60
technoscience, 59, 132
Thompson, Pam, 90
Thornes, John E., 99
Topinka, Robert J., 37
the toxic sublime, 92, 97–98, 101, 106–7
Toxic Substances Control Act (TSCA), 21
Tripp, Katie, 84
Tulloch, John, 124

U.S. Environmental Protection Agency (USEPA), 114
U.S. Fish & Wildlife Service (USFWS), 65–66, 83, 84

Vietnam Veteran's Memorial (VVM), 45
Vintondale, Pennsylvania, 100, 102

Volusia Blue Spring, 67–68, 69, 72, 75, 86
Volusia County, 68, 69, 75

Warm Water Task Force, 84
Washick, Bonnie, 30
Watkins, Carleton, 57, 96
Weldon Spring Site, 8, 106, 107, 115, 116; background of site, 103–4; configuration of site, 112–13, 149, 150; Interpretive Center, 11, 104, 108–9, 111, 113; as a tourist destination, 108, 114, 117
Wells, Justine, 29, 125, 138
Wesselman, Daan, 37–38
West, Tony, 115
West Indian Manatees, 60
West Lake Landfill, 114, 116, 117
wetlands, 78, 135, 148; History Wetlands, 100; at Lake Depue, 144, 145; tidal wetlands of IRL, 75, 76; at Weldon Springs, 90, 103; wetland ecology of Emiquon Preserve, 146
Wilderness Act of 1964, 57
Woods, Barbara, 63–64

Yosemite, 57, 94–95, 96

Zinn, Jens O., 124

About the Author

Dr. Tom Bowers is associate professor in the Department of English at Northern Kentucky University. He has researched and published on various environmental topics including corporate social responsibility, risk communication, mountain top removal, and digital advocacy.

www.ingramcontent.com/pod-product-compliance
Lightning Source LLC
Chambersburg PA
CBHW020123010526
44115CB00008B/950